# The Roadside Geology of Nami[illegible]

by

Gabi Schneider

with contributions by Thomas Becker, Ludi von Bezing, Rolf Emmermann, Steven Frindt, Dougal Jerram, John Kandara, Paul Keller, John Kinahan, Jürgen Kirchner, Volker Lorenz, Volker Petzel, Martin Pickford, Dieter Plöthner, Herbert Roesener, Axel Schmitt, Brigitte Senut, Pete Siegfried, Bernd Teigler, John Ward and Markus Wigand

with 112 figures and 1 table in the text



GEBRÜDER BORNTRAEGER · BERLIN · STUTTGART · 2004

ISBN 3-443-15080–2 / ISSN 0343-737 X

Gedruckt auf alterungsbeständigem Papier nach ISO 9706-1994
Verlag: Gebrüder Borntraeger Verlagsbuchhandlung, Johannesstraße 3 A, D-70176 Stuttgart
E-mail: mail@schweizerbart.de
Internet: http://www.borntraeger-cramer.de
Satz u. Druck: Tutte Druckerei GmbH, Salzweg b. Passau

# Foreword

Since the beginning of the last century, Namibia has attracted scientists from all over the world to study the geology, uniquely exposed in the desert environment. Their research has shaped geological thinking worldwide, and led to the development of many new concepts. Research has been varied and widespread, as demonstrated by the numerous publications on Namibian geology. To name but a few, the work of Hans Cloos provided new insights into the formation of igneous rocks in the 1920s. Research targeting the famous Late Proterozoic Ediacaran fauna, which was first found in Namibia, changed our ideas of the evolution of life on this planet in the 1960s, and the concept of Plate Tectonics was greatly enhanced through studies into the famous Damara Orogen of Namibia in the 1970s and 1980s. More recently, research of glacial sediments within the Damara Orogen has led to the development of the concept of global glaciations, the so-called Snowball Earth concept.

Due to an arid climate and low population density, geological features are ever present and eye-catching in Namibia. The sparse vegetation coupled with a limited number of towns and only isolated settlements leave ample open spaces to observe geological features. It is for these reasons, that both scientists and laymen are attracted to the country, and many a tourist develops a keen interest in geology when touring this beautiful country. This Roadside Geology Guide has been prepared in response to a growing demand for geological information. The idea for a guide was conceived by members of the Geological Survey of Namibia quite some time ago, but it was only when Dr. E. Nägele suggested that a guidebook could be published in the series "Sammlung Geologischer Führer", that the concept was duely developed. We are indebted to him for this suggestion and his support.

The guide provides a general introduction into the geological development of southern Africa and Namibia in particular, followed by general chapters on mineralogy, palaeontology, hydrogeology and mining. Special geological attractions of Namibia are then described in detail. The route descriptions are based on the 1999 simplified 1:2 000 000 geological map of Namibia, published by the Geological Society of Namibia. This simplified map is based on the 1:1 000 000 scale geological map of Namibia which was published by the Geological Survey of Namibia in 1980.

The compilation of this guide has been a team effort, and I have enjoyed the support of many colleagues, who have contributed either as co-authors or by writing entire chapters. Credit for this is given at the beginning of these chapters, but for their contributions I must personnaly thank Dr. T. Becker, L. von Bezing, Prof. Dr. R. Emmermann, Dr. S. Frindt, Dr. D. Jerram, J. Kandara, Prof. Dr. P. Keller, Dr. J. Kinahan, Prof. Dr. J. Kirchner, Prof. Dr. V. Lorenz, V. Petzel, Dr. M. Pickford, Dr. D. Plöthner, H. Roesener, Dr. A. K. Schmitt, Dr B. Senut, Dr. P. Siegfried, Dr. B. Teigler, Dr. J. D. Ward and Dr. M. Wigand.

My colleagues Dr. R. Miller, Prof. Dr. E. Förtsch and D. Hutchins edited and proof-read the manuscript, and made many useful comments for improvement. All diagrams were compiled by S. Cloete and D. Richards, and C. Marais kindly gave permission for the use of her artwork. Secretarial services were provided by E. Grobler. Without the assistance of these people, this book would not have become a reality and I thank all of them very much.

Last, not least, I thank my husband Martin and my children Christian and Kathrin for enduring long journeys throughout Namibia with me, when the only topic up for discussion was the roadside geology of Namibia!

Windhoek, June 2003 Gabi Schneider

# Contents

# 1. Introduction

Namibia, one of Africa's youngest nations, is appropriately named after its Namib Desert, a geological feature that gives the country much of its character, most of its mineral wealth, and a unique flora and fauna. The country borders on the Atlantic Ocean on the west coast of Africa between the $17^{th}$ and $29^{th}$ latitudes. With a total surface area of 824 269 $km^2$, it is the second largest country in southern Africa. Namibia is bounded in the N by Angola and Zambia, in the E by Botswana, and in the S and SE by South Africa. A narrow strip about 300 km long, the Caprivi Strip, extends eastwards up to the Zambesi River, projecting into Zambia and Botswana and touching Zimbabwe.

The subdivision of Namibia into geomorphological units is based on its position on the edge of the African continent and under the influence of the cold Benguela Current. During the Cretaceous and Tertiary, southern Africa completed its separation from the neighbouring parts of Gondwanaland. As a result of isostatic movements, the whole subcontinent underwent several stages of upliftment and the present interior was subjected to erosion. Such isostatic upliftment and the associated erosion are most intense along the edges of the continent, and consequently the Great Escarpment developed. W of the Great Escarpment, the Namib Desert is an 80 to 120 km wide belt, which extends along the entire coastline. The prevailing winds from SW have generated the Namib Sand Sea, a major erg between Lüderitz in the S and Walvis Bay in the N. The other parts of the Namib are rocky desert. Towards the E, the mountainous Central Plateau rises up to 2000 m above sea level and extends over about half of the country. The lower lying northeastern and southeastern areas of the country are extensions of the semi-arid Kalahari Desert. The northern areas beyond the Etosha Pan are bush-covered plains.

The highest mountain in Namibia is the Brandberg in Damaraland (2573 m). Other, well-known mountains are the Gamsberg SW of Windhoek (2347 m), and the Erongo Mountain between Omaruru, Karibib and Usakos (2319 m). The Auas Mountains are situated S of Windhoek. The highest peak in this mountain range, the Moltkeblick, is at 2479 m above sea level the second highest mountain in the country. The Otavi Mountainland E of Otavi reaches 2155 m above sea level. Because of the fact that the southern part of Namibia is so flat, the Brukkaros Mountain (1603 m) near Berseba is a wellkown landmark between Mariental and Keetmanshoop. The highest peak in the Great Karas Mountains in the S is 2202 m high.

Namibia's only perennial rivers lie on her borders, the Orange River in the S, and the Kunene, Kavango, Kwando and Zambesi Rivers in the N. All inland rivers flow sporadically and only after episodical intensive rain showers. After such floods, a river course is often dry for many years. Rivers such as the Hoarusib, Unjab, Ugab, Omaruru, Swakop, Nossob, Kuiseb and Fish usually flow in a good rainy season, however, they may not reach the sea for several years.

According to the Köppen classification, Namibia as a whole has a dry climate characteristic of a desert country. The days are warm to very hot, while the nights are cool. In Windhoek, the average temperature for December is 23,6 °C, and the average maximum 30,6 °C. The average temperature for July is 13,4 °C, the average maximum is 20,4 °C, and the average minimum is 6 °C. The temperatures are influenced by the height of the inland plateau and mountain ranges, and the cold, upwelling Benguela Current, which flows from Antarctica up the W coast of Africa as far as Namibe on the Angolan coast. The Benguela Current is extremely nutrient rich, and has therefore given Namibia one of the richest fishing grounds on earth. Because of its low temperatures, there is little evaporation, and it is therefore responsible for the lack of rain in the Namib Desert. It does, however, generate a thick fog along the coast, which is pushed up to 50 km inland. This life-giving fog supports a wide variety of flora and fauna in the desert.

In the N, October is the hottest month, whereas in the central area December is the hottest and January in the S. The highest temperatures in the country are recorded in the Orange River Valley. During winter, the southeastern portion of the central inland plateau is the coldest. At the coast, the coldest month is August, throughout the rest of the country the lowest temperatures are recorded in July.

Namibia has two rainy seasons: The small rainy season in October and November, and the main rainy season from January to March. Rainfall is erratic and usually occurs in the form of thunderstorms in the afternoon or early evening. The interior plateau has an excessively high evaporation rate as a result of the high day temperatures. The rainfall figures for the Namib Desert are extremely low, from 0 to 50 mm per year.

The southern part of the country has two extreme climatic zones. The southwestern corner bordering Aus and Lüderitz is the only winter rainfall area in Namibia, while the rest of the S, like the rest of the country, receives rain in summer. The summer rainfall area has an average annual rainfall of 100 mm. The shrub savannah and semi-desert vegetation of this region is suitable for karakul farming. The tree-like Aloe, known as the quiver tree, *Aloe dichotoma*, is characteristic of the area. A rich variety of desert succulents are found in the winter rainfall area. About 1200 of Namibia's almost 3500 plant species are found here.

The central part, where the capital Windhoek is situated at an altitude of almost 1800 m above sea level, has an average rainfall of 350 mm per year. The vegetation varies from dwarf shrub savannah in the southern Rehoboth area to highland or mountain thorn savannah around Windhoek, which makes the region suitable for extensive rangeland farming. The area N of Windhoek as far as Tsumeb has an average rainfall of 500 mm per year. This area, which is mopane savannah, thorn bush savannah, mountain savannah and karstveld is suitable for cattle and crop farming.

The Caprivi Strip has the highest annual rainfall in Namibia, ranging from 600 mm in the W to 700 mm at Katima Mulilo on the Zambesi River. Large tree species such as teak for example, occur in abundance here.

# 2. The geological evolution of southern Africa

The geology of Namibia should be seen in the larger context of the geological evolution of southern Africa. This evolution is characterised initially by the formation of cratons as stable parts of the Earth's early crust, surrounded by mobile belts, which were long, relatively narrow, tectonically active areas. Over time, the mobile belts which interconnected areas of already stable crust, became stable themselves, and thereby accreted to form larger cratonic areas. During the course of plate tectonic movements, these cratonic areas combined to form supercontinents which at a later stage from the Proterozoic onwards dispersed again to form new continents. The development of the southern African lithosphere is closely linked to these processes.

The geological evolution of southern Africa spans a period of at least 3500 Ma. The development of cratonic areas during the Archaean and Palaeoproterozoic was followed by the assembly of the supercontinent Rodinia some 1000 Ma ago. The dispersion of Rodinia followed and preceeded the assembly of the supercontinent Gondwanaland about 550 Ma ago. Southern Africa formed as part of Gondwanaland until break-up some 180–130 Ma ago, since when stable geological conditions have prevailed as southern Africa developed as a part of the African Continent (Fig. 2.1) (Truswell 1977).

In various parts of the World, the first continental crust formed Archaean cratons between 4000 and 2500 Ma ago. In southern Africa, two such Archean cratons occur, namely the Kaapvaal Craton and the Zimbabwe Craton (Fig. 2.2). The Kaapvaal Craton stabilised about 3000 Ma ago, and consists of sediments resting on granites, gneisses and deformed greenstones. In the Zimbabwe Craton granites and greenstones are more prominent, and sediments occur only along the eastern and northwestern margins. The ages of the granites vary between 3500 and 2600 Ma, and whilst the accumulation of sediments had already occurred on the Kaap-

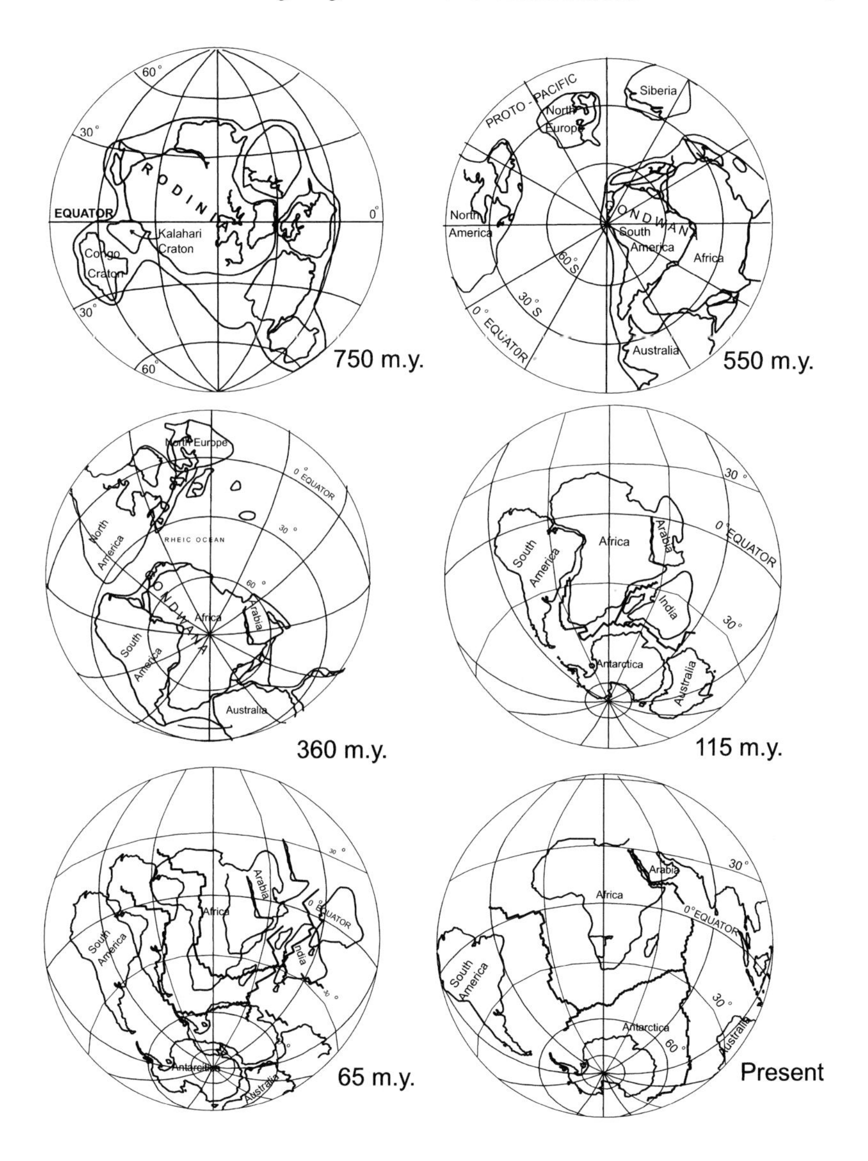


**Fig. 2.1:** Africa's position and plate tectonic movements since 750 Ma before present.

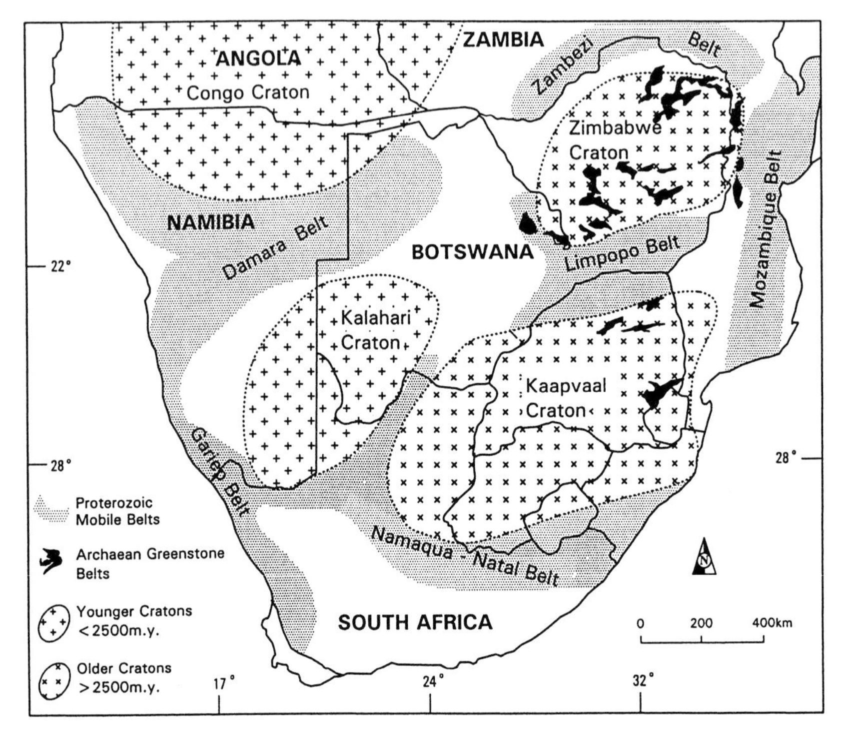


**Fig. 2.2:** Cratons and mobile belts in southern Africa (after Annhaeusser & Button 1974).

vaal Craton, volcanism and granite intrusion continued in the Zimbabwe Craton, to such an extend that it only became a stable part of the Earth's crust 400 Ma later.

In the northwestern part of southern Africa, the delineation of Archean Cratons is more difficult, since the area is largely covered by younger rocks. However, some gneisses in Angola have been dated at 3500 Ma, and are interpreted as the first nucleus of the Angola Craton.

Following the establishment of these stable areas in the Earth's crust, sediments were deposited adjacent to the cratons. These sediments were later metamorphosed and intruded by volcanic rocks, forming mobile belts between the cratons. The Kaapvaal and Zimbabwe Cratons are separated by the Limpopo Mobile Belt, with a long continuing sequence of tectonic and magmatic events, spanning a time between 3800 and 2700 Ma ago.

In the Palaeoproterozoic and Mesoproterozoic between 1700 and 1300 Ma ago, thick sequences of sediments and volcanics were deposited in several small basins between the Kaapvaal and Zimbabwe Cratons to the SE and between the Kaapvaal and the Angola Cratons in the N. These sediments and volcanics are today preserved as basement inliers. The accretion of continental crust in southern Africa continued further with the formation of the Namaqua-Natal Belt on the southern and southwestern margin of the Kaapvaal Craton between 1300 and 1000 Ma ago. Through this process stabilised landmasses were connected to form the Proterozoic Congo and Kalahari Cratons, which essentially are extensions of the Archaean Angola and Kaapvaal Cratons respectively (Trompette 1991).

During the Neoproterozoic and Early Palaeozoic between 750 and 450 Ma ago major metamorphism affected large parts of the entire African continent. This, the Pan-African metamorphic event, not only deformed upper Precambrian sedimentary sequences deposited along the margins of the older cratons, but also older rocks. In southern Africa, it produced the important Damara Belt, the Gariep Belt, the Mocambique Belt and the Zambesi Belt.

The origin of the southern African Precambrian mobile belts is not exactly clear. However, geological patterns suggest a plate tectonic model for the Pan-African mobile belts, and the formation of the Damara Belt, for example, is interpreted as an active continental margin, with the plate containing the Kalahari Craton being subducted underneath the Congo Craton (Hunter & Pretorius 1981).

Once Gondwanaland had been established about 500 Ma ago, depression of the continental crust in large parts of southern Africa and South America behind the volcanic Andean mountain chain permitted the deposition of huge amounts of sediments. The Karoo Sequence covering major parts of southern Africa dates from the late Carboniferous to the early Jurassic between 300 and 180 Ma ago. Karoo sedimentation began with the deposition of the Dwyka tillites during a Permo-Carboniferous glaciation. Many of the later Permian and Triassic sediments are of fluvial origin, and the sedimentary cycle culminated under arid conditions with the deposition of aeolian sandstones (Truswell 1971).

About 150 Ma ago Africa started to break away from the remainder of Gondwanaland and by about 125 Ma ago, in the early Cretaceous, Africa had emerged as a separate plate. Just prior to the rifting process, large volumes of volcanic melt were erupted on surface and intruded the crust to form dykes, and marine sediments were deposited in the developing basins.

The most recent deposition of marine sediments throughout the Cenozoic occurred on the shelf and in the coastal areas surrounding southern Africa. Along the arid west coast, the aeolian sand sea of the Namib Desert formed. The geological history of the southern African interior since the break-up of Gondwanaland is largely one of erosion. The inland basin of the Kalahari formed and has been filled with terrestrial sediments during most of the Cenozoic (Truswell 1971).

# 3. Namibia from the Archaean to the Cenozoic

Namibia's varied geology encompasses rocks of Archaean to Quaternary age, thus covering more than 2600 Ma of Earth's history. About half of the country's surface is bedrock exposure, while the remainder is covered by surficial deposits of the Kalahari and Namib Deserts. There are five main rock-forming periods (Tab. 3.1).

The oldest rocks occur within metamorphic complexes of Vaalian (> 2000 Ma) to Early Mokolian (2000 to 1800 Ma) age. These form a basement to the rocks of the next major period, which include sedimentary and volcanic successions, of Middle to Late Mokolian (1800 to 1000 Ma) age. This period involved the formation of the Rehoboth-Sinclair magmatic arc and the Namaqualand Complex. The final events in this evolution led to the assembly of the supercontinent Rodinia.

The third main event, the formation of the Damaran mountain belt, started with intracontinental rifting and sedimentation about 900 Ma ago and culminated in continental collision during the Damara Orogeny between 650 and 450 Ma ago. This coincides with the assembly of the Gondwanaland supercontinent.

Extensive erosion of the Damaran mountainbelt preceded the fourth phase, the deposition of the Karoo Sequence between the Permian and the Jurassic, some 300 to 135 Ma ago. Extensive volcanism accompanied the break-up of Gondwanaland at the end of this phase. Today, Cretaceous to recent deposits cover many of the older rocks and constitute the fifth phase (compare 2. and Fig. 2.1) (Miller 1992).

## Vaalian to Early Mokolian rocks (> 1800 Ma)

Metamorphic Complexes crop out in several major inliers in northern and central Namibia (Fig. 3.1), but having undergone several cycles of metamorphism, the reconstruction of their depositional environment is quite difficult. They can, however, be interpreted as part of the developing Proterozoic Congo and Kalahari Cratons.

**Tab. 3.1:** Correlation of the Namibian Stratigraphy.

<table>
<tr><th colspan="2">International Geological Time-Scale Subdivisions</th><th>Southern African Subdivisions</th><th>Namibian Stratigraphic Units</th></tr>
<tr><td>Cenozoic<br>< 65 Ma</td><td rowspan="2">Quaternary to Cretaceous<br>< 135</td><td rowspan="2">< 135 Ma</td><td rowspan="4">Namib Desert<br>Kalahari Sequence<br>Etendeka Group<br>Post-Karoo<br>Complexes<br>Karoo Sequence</td></tr>
<tr><td rowspan="2">Mesozoic<br>25 – 65 Ma</td></tr>
<tr><td rowspan="2">Permo-Carboniferous to Jurassic<br>300 – 135 Ma</td><td rowspan="2">300 – 135 Ma</td></tr>
<tr><td rowspan="2">Paleozoic<br>540 – 250 Ma</td></tr>
<tr><td>Ordovician to Cambrian<br>540 – 460 Ma</td><td>Ordovician to Cambrian<br>540 – 460 Ma</td><td>Damara Granites</td></tr>
<tr><td rowspan="4">Precambrian > 540 Ma</td><td>Neoproterozoic<br>1000 – 540 Ma</td><td>Namibian<br>1000 – 540 Ma</td><td>Nama Group<br>Damara Sequence<br>Gariep Complex</td></tr>
<tr><td>Mesoproterozoic<br>1600 – 1000 Ma</td><td rowspan="2">Middle to Late Mokolian<br>1800 – 1000 Ma</td><td rowspan="2">Gamsberg Granite<br>Sinclair Sequence<br>Namaqualand Complex<br>Rehoboth Sequence<br>Fransfontein Suite</td></tr>
<tr><td rowspan="2">Paleoproterozoic<br>2500 – 1600 Ma<br>and Archaean<br>> 2500 Ma</td></tr>
<tr><td>Early Mokolian<br>2000 – 1800 Ma<br>to Vaalian > 2000 Ma</td><td>Kunene Complex<br>Elim Formation<br>Khoabendus Group<br>Mooirivier Complex<br>Neuhof Formation<br>Hohewarte Complex<br>Abbabis Complex<br>Grootfontein Complex<br>Huab Complex<br>Vioolsdrif Suite<br>Orange River Group<br>Epupa Complex</td></tr>
</table>

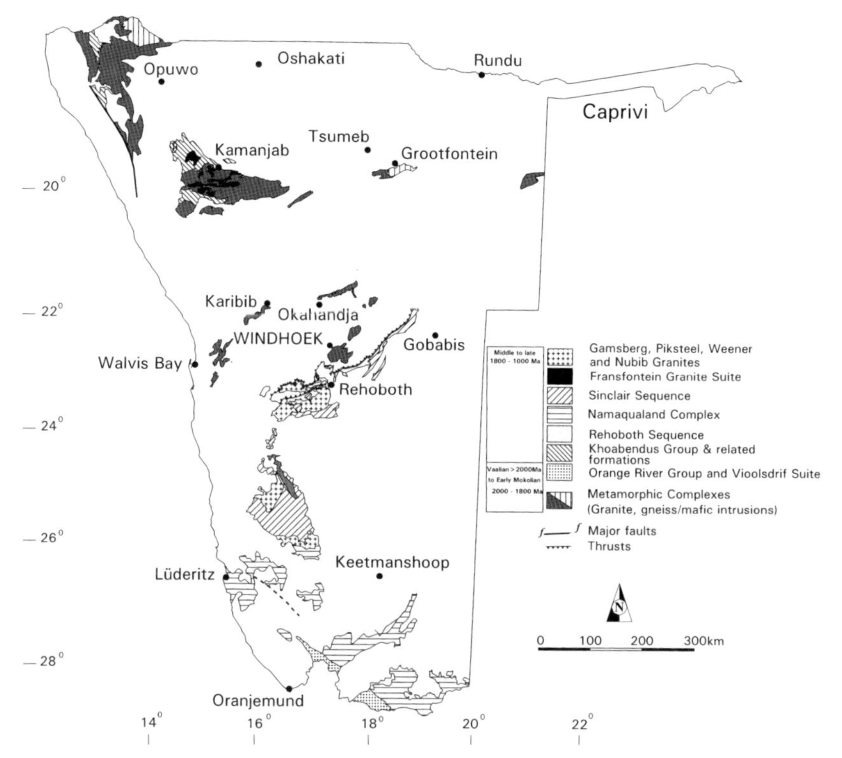


**Fig. 3.1:** Pre-Damara geology of Namibia.

The oldest rocks in Namibia are gneisses exposed in the Hoarusib Valley NW of Sesfontein in northwestern Namibia. They have been dated at 2645 Ma (Seth et al. 1998) and might be related to the gneisses of the Epupa Complex which straddles the northern border with Angola. The Epupa Complex is intruded by anorthosites of the Kunene Complex of uncertain age, and although occurring mainly in Angola, its southern tip extends into Namibia. The main rock type is massive anorthosite with intercalated troctolite and norite. Dykes of syenite, granite and carbonatite cut the Kunene Complex, as well as the enveloping Epupa Complex.

Further S, light gneisses containing layers of quartzite, mica schist and amphibolite are the main rock types of the Huab Complex W of Fransfontein, whilst coarse-grained granite which is cut by amphibolite dykes makes up the Grootfontein Complex in the NE. A large, but poorly exposed basic intrusive body belonging to the same complex occurs E of Grootfontein.

Rocks of the Khoabendus Group occur in two separate regions in the northwestern and southwestern parts of the Kamanjab Inlier. The Group is composed largely of a volcanic lower portion and a sedimentary upper portion. The volcanics comprise acid pyroclastics, feldspar porphyry, dacite and rhyolite. The sediments include quartzite, conglomerate, limestone, dolomite, chert and banded iron formation.

In the Erongo Region, various gneisses, amphibolites and pegmatites constitute the Abbabis Complex of central Namibia. E of Walvis Bay, banded gneiss is the main rock type, while marble, conglomerate and calc-silicate also occur and in the Karibib-Usakos area, schist, conglomerate, quartzite and minor limestone are found. SW of Karibib there is a thick succession of arkose with intercalated quartzite and conglomerate bands. W of Okahandja, the Abbabis Complex is represented by gneiss, quartzite, calc-silicate and marble, while to the E of that town it largely consists of reddish gneiss. Plutons of granodiorite of the Abbabis Complex are found S of Karibib.

The Hohewarte Complex in the Rehoboth area consists of a partly migmatised succession of schist, quartzite and amphibolite, associated with some granite. Migmatitic gneisses and amphibolites with intensive intrusive granite also make up the Mooirivier Complex.The Neuhof Formation overlies the rocks of the Mooirivier Complex, and consists of a succession of volcanic rocks, quartzite, conglomerate, schist and calc-silicate with intrusive granite. A white basal quartzite overlain by amphibolite, schist, quartzlite, limestone and rhyolite make up the Elim Formation W of Rehoboth.

In southern Namibia, intermittent outcrops of the Orange River Group occur in a belt N of the Orange River. The Group is made up largely of andesitic calc-alkaline volcanic rocks with only minor sediments. The Vioolsdrif Intrusive Suite extensively intrudes the Orange River Group, consisting mainly of granodiorite and adamellite, with minor gabbro and diorite also present.

**Middle to Late Mokolian rocks (1800 to 1000 Ma)**

Erosion of the old cratons is responsible for the accumulation of sediments of this age along their fringes. These sediments were metamorphosed during the Kibaran Orogeny between 1400 and 900 Ma ago, forming mobile belts interconnecting the old cratons which finally led to the assembly of the supercontinent of Rodinia (compare 2.)

The 1800 Ma old Rehoboth Sequence (Fig. 3.1) unconformably overlies the Elim Formation in the central part of the country and is thought to have formed in the back-arc basin of a magmatic arc, which possibly extended from present-day South America through Namibia, Zambia, Zimbabwe and Tanzania to Uganda. Closure of this basin and continental collision resulted in a major phase of deformation and metamorphism accompanied by the intrusion of granitic and basic rocks ranging from 1670 to 1420 Ma in age. The Rehoboth Sequence consists of the Marienhof Formation at the base, which is overlain unconformably by the Billstein Formation, with the Gaub Valley Formation occurring further W. The Marienhof Formation consists of alternating layers of quartzite, phyllite, conglomerate and volcanic rocks. The Billstein Formation is dominated by phyllite and quartzite, with some volcanic rocks occurring as intercalations. Characteristic brown quartzite is the main rock type of the Gaub Valley Formation, with quartzite, phyllite, conglomerate, limestone and some volcanic rocks also occurring.

The Namaqualand Complex (Fig. 3.1) covers extensive areas in the S of Namibia. It can be subdivided into pre-tectonic meta-sediments, gneisses, amphibolites, charnockites, gabbros and ultrabasic rocks; syn-tectonic granites and late- to post-tectonic granites. The pre-tectonic meta-sediments accumulated along the fringes of the Kaapvaal Craton and were metamorphosed about 1200 Ma ago. Concomitantly, a magmatic arc formed, and the magmatic activity lasted for about 200 Ma.

Rocks of the Sinclair Sequence (Fig. 3.1) accumulated along an active continental margin in the Helmeringhausen area and in a back arc region and intracontinental rifts further N. Deposition of the sediments took place in narrow-fault-bounded troughs aligned around the Kalahari Craton in the Helmeringhausen-Solitaire area during three broad cycles of volcanism, plutonism and sedimentation. Volcanic rocks at the base are intercalated with sandstones and shales. The intrusion of the Kotzerus granite was followed by the deposition of conglomerates, sandstones and shales with interbedded lava of the Kunjas Formation. The second cycle started with the deposition of various volcanics and interbedded sandstone, conglomerate and shale of the Barby Formation. These rocks were intruded by gabbro, syenite and granite, whereafter more sediments and lava were deposited. The youngest granites have been dated at 1000 Ma. The third cycle is represented by swarms of acid and basic dykes, quartz porphyry and minor granite, followed by the deposition of red sandstones, conglomerates and shales of the Aubures, Doornpoort and Eskadron Formations.

The Fransfontein Granitic Suite intruded rocks of the Huab Complex and the Khoabendus Group. Granodiorite and granite are dominant, small diorite bodies are also present. The Gamsberg, Piksteel, Weener and Nubib Granites extensively intruded the Mokolian rocks in the central part of the country (Fig. 3.1).

**Namibian to Early Cambrian (1000 to 500 Ma)**

The Neoproterozoic Damara Orogen and Gariep Belt were formed during a complete plate tectonic cycle in successive phases of intra-continental rifting, spreading, the formation of passive continental margins, mid-ocean ridge development and subsequent subduction and continental collision involving the Congo and Kalahari Cratons, and culminating in the formation of Alpine-type mountain belts. The coastal and intracontinental arms of the Damara Orogen underlie much of NW and central Namibia. Along the southwestern coast, the Gariep Belt represents the southern extension of the Damara Orogen (Fig. 3.2, Tab. 3.2).

The basal Nosib Group was laid down in or marginal to intracontinental rifts and consists of quartzite, arkose, conglomerate, phyllite, calc-silicate and subordinate limestone. Local ignimbrites with associated subvolcanic intrusions range from 820 to 730 Ma in age (Fig. 3.3).

Continental break-up followed the initial rifting and the widespread clastic and carbonate deposition of the Kudis, Ugab and lower Khomas Subgroups of the Swakop Group took place in the marginal shelf regions and extended far beyond the shoulders of the early rift graben. Interbedded schists, banded iron formation and basic lavas point to quite variable depositional conditions. The Chuos Formation of the lower Khomas Subgroup forms the most important marker, since it occurs throughout the Damara Orogen. It comprises a variety of rock types ranging from mixtites deposit-

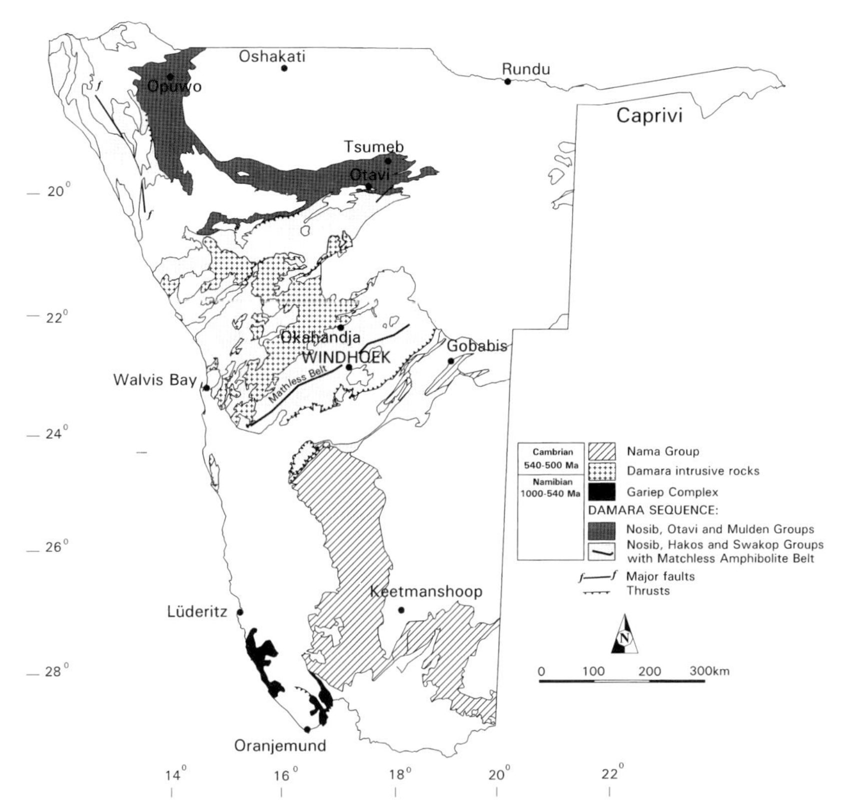


**Fig. 3.2:** Damara-age geology of Namibia.

ed by glaciers and schist to conglomerate and quartzite (Fig. 3.4). Similar rocks were deposited worldwide some 700 Ma ago and are regarded vestiges of several phases of global glaciation (Hoffman et al. 1998a, 1998b).

To the N, a stable platform developed, where the massive carbonates of the Otavi Group were deposited. At the same time, deep water sediments of the Auas and Karibib Formations were deposited on either side of a narrow ocean separating the Kalahari and Congo Cratons. Subsequently, thick layers of greywacke and fine-grained rocks of the Kuiseb Formation were

**Tab. 3.2:** Stratigraphy of the Damara Orogen.

| Ma | Group | | Subgroup | | Formation | |
|---|---|---|---|---|---|---|
| 535 | Nama<br>Mulden | | | | | |
| 650<br>700 | Swakop | Otavi | Khomas | Tsumeb | Kuiseb<br>Karibib<br>Auas<br>Chuos | |
| 730 | | | Ugab<br>Kudis | Abenab | | |
| 820 | Nosib | | | | | |

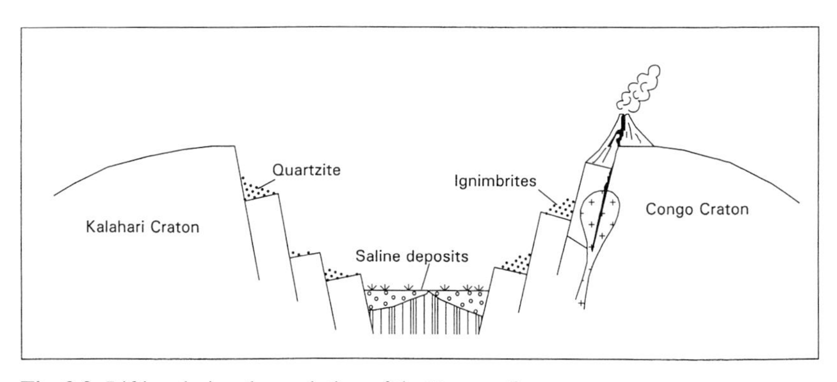


**Fig. 3.3:** Rifting during the evolution of the Damara Orogen.

deposited in that ocean. A mid-ocean ridge developed, where submarine basalts were erupted (Fig. 3.5).

Thereafter, a reversal of the extensional movements led to subduction of the oceanic crust under the Congo Craton and eventually to continental collision (Fig. 3.6). The associated deformation created the Damara moun-

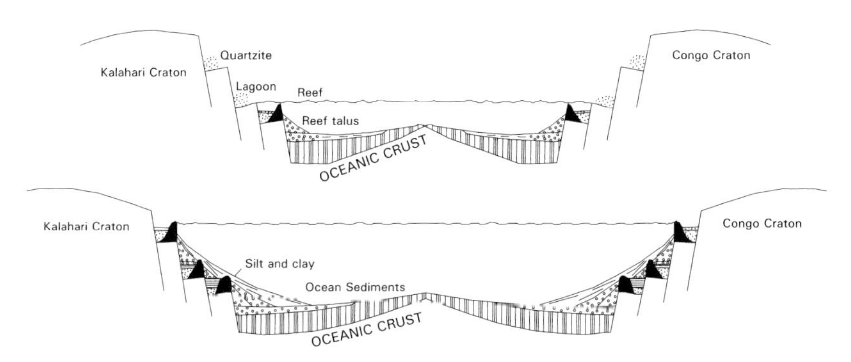


**Fig. 3.4:** Continental break-up during the evolution of the Damara Orogen.

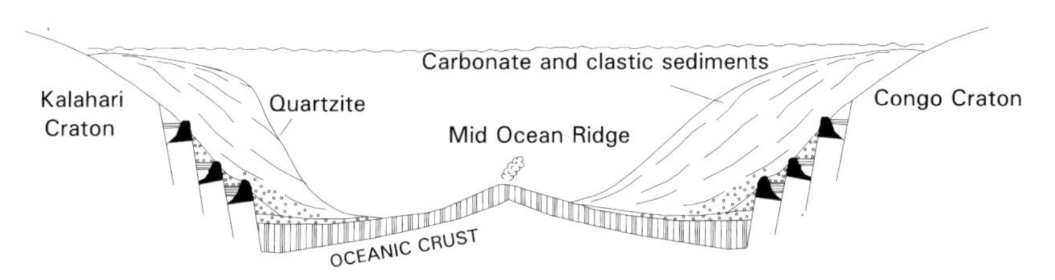


**Fig. 3.5:** Spreading and formation of a mid-oceanic ridge during the evolution of the Damara Orogen.

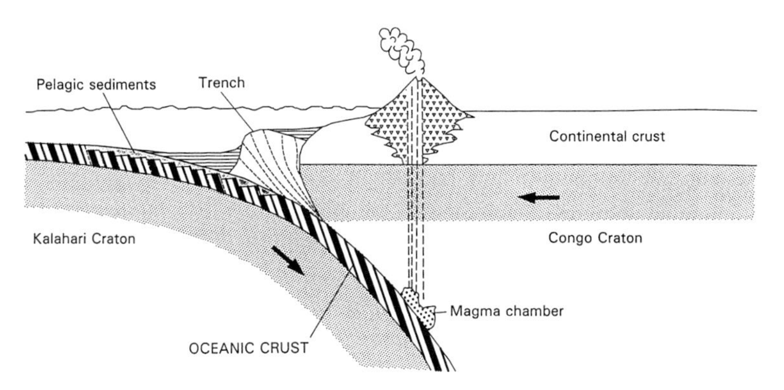


**Fig. 3.6:** Continental collision during the evolution of the Damara Orogen.

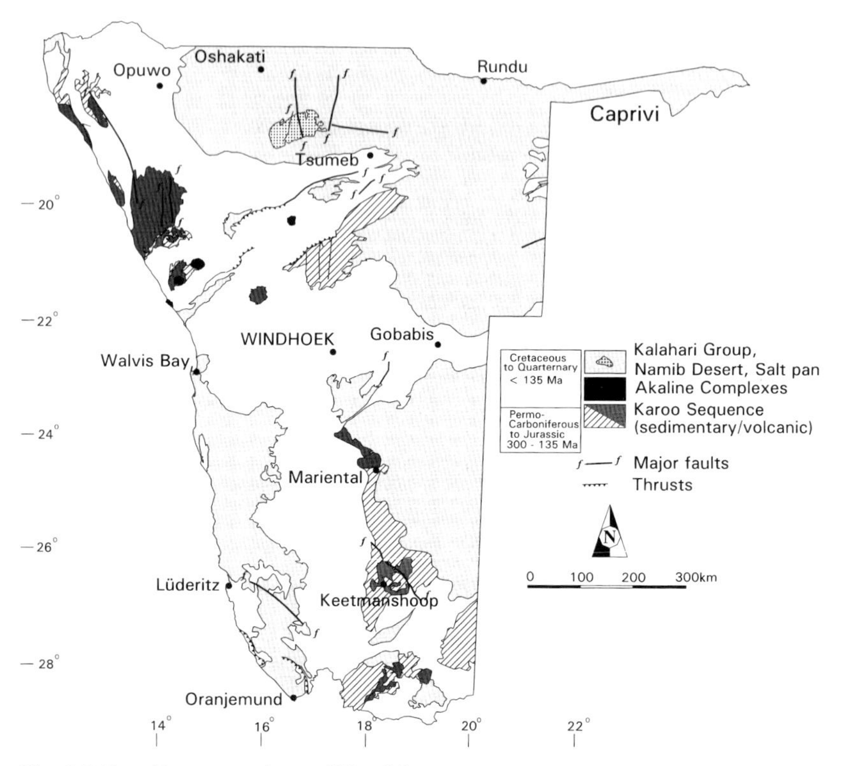


**Fig. 3.7:** Post-Damara geology of Namibia.

tain belt. The intracontinental arm of the Damara Orogen developed paired metamorphic belts with a high temperature – low pressure zone in the NW and a low temperature – high pressure zone in the SE. Central Namibia was the location of an active continental margin with a northwestwards dipping subduction zone (Fig. 3.6). The zone immediately to the SE became a trench zone, where Kuiseb Formation rocks underwent metamorphism to form meta-greywackes and schists. The basalts of the mid-ocean ridge were also metamorphosed and today occur as the 350 km long Matchless Belt of amphibolites. In the coastal arm, another set of paired metamorphic belts developed. The intrusion of syn- to post-tectonic granites and pegmatites accompanied the mountain-building process between 650 and 460 Ma ago.

Erosion of the young mountain belt started some 650 Ma ago and led to the deposition of a northern molasse, the Mulden Group. Deposition of the Nama Group in a shallow syn-tectonic foreland basin resting on a stable platform commenced 600 Ma ago and lasted for about 60 Ma. It covers much of southeastern Namibia. Basal quartzites are overlain by limestones, shales and sandstones.

On the basis of stratigraphy, structure, grade of metamorphism, plutonic rocks and geochronology, the intracontinental branch of the Damara Orogen has been subdivided into several zones. From N to S they are the Northern Platform, the Northern Zone, the Central Zone, the Okahandja Lineament Zone, the Southern Zone, the Southern Margin Zone and the Southern Foreland. The Northern Platform contains the thick platform carbonates of the Otavi Group overlain by the molasse of the Mulden Group. The Northern Zone represents a transition between the platform carbonates and the shelf sediments of the Swakop Group. The Central Zone (Swakop Zone) is the high temperature – low pressure zone of the orogen, and characterised by numerous granitic plutons.

The Okahandja Lineament is one of the most important boundaries in the orogen, where major changes in the stratigraphic succession, structural style and age of deformation occur. It marks the southern edge of the Central Zone. Like the Southern Zone (Khomas Zone), the Okahandja Lineament Zone comprises the Kuiseb Formation, but with different metamorphic features. The Southern Margin Zone (Hakos Zone) is a zone of intense thrusting which occurred during continental collision. The Southern Foreland comprises foreland and molasse sediments of the Nama Group. In addition, the Naukluft Nappe Complex was emplaced from the Southern Margin Zone onto the Southern Foreland. The coastal branch has been subdivided into the Western, Central, Eastern and Southern Kaoko Zones.

In the S, the Gariep Complex is subdivided into an eastern and a western part. In the eastern Port Nolloth Terrane, quartzite, arkose, argillite, carbonate, conglomerate, felsite and greenstone of the Rosh Pinah Formation are overlain by meta-sediment and limestone of the Hilda Formation, and meta-sediment, dolomite and iron formation of the Numees Formation. In the western Marmora Terrane, basic lava, dolomite and schist of the Grootderm Suite are overlain by quartzite, dolomite and iron formation of the Oranjemund Suite and conglomerate, dolomite, arkose, quartzite, shale and limestone of the Bogenfels Formation. The Complex was intruded by granite between 500 and 600 Ma ago.

**Permo-Carboniferous to Jurassic (300 to 135 Ma)**

Following the assembly of Gondwanaland towards the end of the Damara Orogeny, stable continental conditions prevailed throughout the Palaeozoic and and the early to middle Mesozoic. By the end of the Carboniferous, 200 Ma of erosion had left only remnants of the Damara and Gariep mountainbelts. Most parts of present-day Namibia had become vast peneplains, separated only by those remnants. The peneplains were to become the sedimentary basins in which the rocks of the Permo-Carboniferous to Jurassic Karoo Sequence were to be deposited. The Aranos, Huab, Waterberg and Owambo Basins occur in the southeastern and northerly parts of Namibia (Fig. 3.7). Within Gondwanaland, southern Africa initially occupied a position close to the south pole (Fig. 2.1), and a huge ice sheet covered the region. Basal glaciogenic rocks of the Permo-Carboniferous Dwyka Formation include moraines and fluvio-glacial as well as glacio-marine deposits. Glacial pavements are preserved in the S and NW, and deeply incised, westerly flowing glacial valleys, are present in the Kaokoveld, where for example the Kunene Valley in the area of the Baines Mountains is over 1500 m deep.The Dwyka glaciation ended approximately 280 Ma ago, when plate tectonic movements brought southern Africa to a more moderate climatic realm.

At the end of the Dwyka glaciation, the melting ice sheet provided ample water to create an environment with huge lakes and rivers. Consequently, the Dwyka Formation is overlain by lacustrine grey- to green-weathering shales, mudstones, limestones, sandstones and coal-bearing shales of the Prince Albert Formation, which in turn is overlain by fluviatile sandstones. Mid-Permian rocks in the Aranos Basin consist of 600 meters of shales with thin limestone layers overlain by shale and sandstone. In the Huab basin, mid-Permian rocks are represented by purple shales and sandstones. Red, Triassic conglomerates, mudstones, sandstones and grits up to 600 meters thick make up the Omingonde Formation of the Waterberg Basin.

Further severe changes of the environmental conditions about 200 Ma ago led to the establishment of an extremely arid climate. A huge desert formed and aeolian deposition prevailed. The Triassic to Jurassic Etjo Sandstone Formation is a remnant of this desert and comprises fossil sand dunes. It overlies the middle Permian rocks in the Waterberg and Huab Basins.

In the Mariental area, about 360 meters of the 180 Ma old basalts of the Kalkrand Formation overlie Dwyka beds. Basalts in the Kavango and Caprivi areas may be of similar age. The eruption of these basalts can be seen in connection with the early break-up of Gondwanaland, when southern Africa started to separate from the land towards the E (Fig. 2.1). Volcanic rocks of the Etendeka Formation in the NW have an age of about 132 Ma, and mark the separation of Africa from South America. Extensive dolerite sills and dyke swarms are related to the volcanic rocks.

**Cretaceous to Quaternary (< 135 Ma)**

Complex intrusions with ages of 137 to 130 Ma occur in a zone extending from the coast N of Swakopmund in a northeasterly direction. Some of them are extremely complex layered intrusions and contain rhyolite, granophyre, granite, syenite, foyaite, gabbro, dunite, pyroxenite and carbonatite. They are not related to any orogeny and interpretated as a result of a hot mantle plume. Over 60 kimberlite pipes and related rocks have been recognised and are presumed to be of Cretaceous age. Most of them occur in the Gibeon area.

During the Cretaceous and Tertiary, southern Africa completed its separation from the neighbouring parts of Gondwanaland (Fig. 2.1). As a result of isostatic movements, the subcontinent underwent various stages of uplift and the present interior was subjected to erosion. Such isostatic uplift is most prominently visible along the edges of a continent, where erosion is most intense. Consequently, the entire southern Africa is surrounded by the Great Escarpment, and a sedimentary basin, the Kalahari Basin, has formed in the center of the subcontinent. Although the main sediment transport has occurred towards the Atlantic Ocean in the W, some transport in an easterly direction has produced the conglomerates and sands of the Kalahari Sequence, which forms an extensive cover of terrestrial origin in the eastern and northern parts of the country.

Through deep erosion from the end of the Cretaceous about 65 Ma ago onwards, the Great Escarpment moved eastwards, leaving behind the peneplain today occupied by the Namib Desert. Consequently, the succession in the Namib Desert consists of local Cretaceous deposits and extensive terrestrial Tertiary to recent sediments. By the end of the Cretaceous, the Orange River was already a major drainage system and deposited vast amounts of sediments in a huge submarine offshore delta.

The Pleistocene glaciation in the northern hemisphere, and the associated melting of the arctic ice sheet at the end of the glaciation also resulted in climatic variations and sea level changes in the southern hemisphere. The strong long-shore drift transported much of the sediment from the offshore delta northwards to form gravel- and sandy beaches at various sea levels. The sand of these beaches has been picked up by the wind and blown inland to form the sand sea and dune belts of the Namib Desert. Most of the large westerly flowing rivers have a Tertiary to recent fill consisting of fluviatile sediments as well as thick, calcrete-cemented sand and conglomerate.

# 4. An overview of the Namibian mineralogy

by Paul Keller and Ludi von Bezing

Namibia is blessed with minerals of economic importance like diamonds, semi-precious stones and a variety of ore minerals. Many minerals of scientific interest or just beautiful crystals have been found, often incidental to prospecting and mining activities. Minerals from Tsumeb for example have graced the collections of museums throughout the world for about a century and hundreds of geologists, mineralogists, curators and collectors visit Namibia annually. Tsumeb with well over fifty species new to science and over 250 different minerals, ranks as the most important locality in Africa and one of the premier mineral localities in the world (Wilson 1977, Keller 1984, Gebhard 1991). The distribution of known mineral-producing localities is not surprisingly biased in favour of the Otavi Mountainland where Tsumeb is located and intensive exploration in the region has located over 600 deposits and prospects. There are however thousands of other deposits, particularly a large number of pegmatites in central and southern Namibia, that are known for their minerals. Thus Namibia ranks on the African continent second only to the Democratic Republic of Congo concerning the number of new mineral species discovered (Daltry 1992).

The mineral deposits can be grouped according to a relatively simple scheme into a classification based on their genesis. Hydrothermal deposits often have a simple mineralogy. The exception is Tsumeb, where about 50 different primary ore minerals occur. Outstanding specimens are, in addition to large and well developed crystals of tennantite, the germanium- and gallium-minerals germanite, renierite, gallite and briartite (Lombaard et al. 1986). The diversity of minerals found in the famous Tsumeb deposit, where a gossan and a second large deep-seated oxidation zone are developed, is however due to the occurrence of these large oxidation zones. The large variety in both, the conditions of formation and the chemical composition of the primary ore is responsible for the uniqueness of Tsumeb (Keller 1977, Ingwersen 1990). Many of the minerals from Tsumeb occur only here or are the best of their species. Azurite, dioptase, cerrussite, smithsonite, wulfenite, mimetite, anglesite, carminite, mottramite, duftite, tsumcorite and pseudomorphs of malachite after azurite are the best known specimens of these minerals. Studies of some of the other minerals have helped to understand crystal chemistry, especially copper-zinc arsenates as well as the large number of other arsenides and lead silicates.

The Kombat deposits are also of hydrothermal origin, but a metamorphic overprint with very special physico-chemical conditions has resulted in an unusual mineralogy (Innes & Chaplin 1986). Lenses of manganese and iron ores, which are associated with Cu-Pb-Ag-sulphides show some similarities with Langbån, Sweden and Franklin, New Jersey. The large, intensely red nambulite crystals initially made Kombat famous. However, later a large number of new minerals were discovered from the manganese and iron ores, such as kombatite, asisite, johninnesite and amaraite.

The Mississippi Valley type lead-zinc deposits of Berg Aukas and Abenab belong to a third notable group of mineral deposits in the Otavi Mountainland (Killik 1986). They are known for minerals such as descloizite, smithsonite and willemite, which form beautiful crystals in surface enrichments and karst cavities. Berg Aukas in particular is world renowned for its fine descloizite and Abenab has produced the largest crystal of vanadinite (with a length of 12 cm) in the world (Rickwood 1981).

Unusual minerals such as norsethite and benstonite occur in the exhalative lead-zinc deposit at Rosh Pinah, near the Orange River.

The best known hydrothermal copper deposit is the Onganja Mine 80 km NE of Windhoek, which is famous for the largest known euhedral cuprite crystals in the world measuring 14 cm in length and width and weighing over 2 kg. High quality dioptase is mined from supergene veins at Omaue, Kaokoland (Jacob 1987) although many hydrothermal veins carry only microminerals, like the vanadinite from Namib Lead Mine, 30 km NE of Swakopmund, quite a few have yielded new mineral species such as namibite from the polymetallic hydrothermal quartz veins near Khorixas, or idaite, from the Ida Mine in the Khan River Valley E of Swakopmund.

Chalcedony, formed by low-temperature hydrothermal activity is found at several places, and blue chalcedony of high quality is found at Otjoruharui 251 NE of Okahandja. Lead mineralisation in one part of the deposit led to a number of secondary lead minerals of which orange-red crocoite is the most notable. Chalcedony also occurs near the Rössing Mountain. Veins of siderite and fine bluish banded agate, that are mined at

Ysterputz 254 SW of Karasburg (Schneider & Seeger 1992) are also probably of hydrothermal origin.

Pegmatites in southern Namibia belong to the Namaqua Metamorphic Complex which crops out along the southern border of Namibia with South Africa and has an age of 900 to 1000 Ma. Pegmatites in the central and northern parts of the country are part of the Damara Orogen and intruded during late Pan-African times between 510 and 480 Ma ago. They form a distinct belt striking in a SW-NE direction.

The pegmatites of economic interest which have been mined in Namibia belong to the group of rare-metal pegmatites. They formed at intermediate depth and are often typically zoned. Such a zonation is especially well exposed in some pegmatites of the Tantalite Valley in the Karasburg District. Pegmatites in the Karibib area have been extensively researched, like for example Tsaobismund (Uebel 1977), Brabant (Rose 1981), Okatjimukuju (Keller & von Knorring 1989) and Rubicon (Diehl & Schneider 1990). The Namibian rare-metal pegmatites can be subdivided according to their mineralisation of beryllium, niobium/tantalum, lithium, tin and semi-precious stones, many of which contain excellent specimens.

The tin-tantalite pegmatites of the Northern and Central Tin Belts are of high commercial but low mineralogical interest. Exceptions are Tsomtsaub, where good crysoberyl crystals and the scarce mineral nigerite occur and a pegmatite near Mile 72 in the National West Coast Recreation Area, where blue jeremejevite of gem quality has been recovered from.

The lithium pegmatites offer a much more interesting mineralogy. Whereas lepidolite, petalite and amblygonite-montebrasite are the usual lithium minerals in the Karibib Pegmatite District, for example at Rubicon and Helikon, Etiro 50, Daheim 106 and Karlsbrunn 42, spodumene is mainly found in the Karasburg District. Spodumen has also been mined from De Rust, NE of Brandberg, from a few pegmatites of the Northern Tin Belt and at Donkerhuk 91 and Etusis 75. The rare lithium mineral eucryptite has been found in large nodules in the Strathmore Tin Mine, Northern Tin Belt.

Tantalite-columbite, beryl, tourmaline, pollucite, topaz, microlite, apatite and native bismuth are accessories of the lithium pegmatites. These minerals together with the rock-forming pegmatite minerals occasionally form beautiful crystals with well developed faces. Shiny quartz crystals, which are partly of the citrine variety, and more than 1 m in length, have been found at Rubicon. At this locality, orthoclase, water-clear or bluish albite, yellowish green muscovite, bluish amblygonite, native bismuth up to 15 cm in size, large tantalite and bluish and white apatite crystals have also been found.

Phosphate-bearing pegmatites are another source of a large number of minerals. They belong mainly to the Karibib Pegmatite District and the Nainais-Kohero Pegmatite Belt. About 80 primary and secondary phosphate minerals are known. New mineral species are ernstite, karibibite, brabantite and giniite, and particularly exceptional is the frequency and size of kryzhanovskite-landesite, eosphorite-childrenite, arrojadite, and augelite. Although the phosphates are mostly of microscopic size, they are important in understanding the genesis of pegmatites (Keller 1991, Keller et al. 1994). Outstanding occurrences are Sandamap, 40 km NW of Usakos, and Davib Ost 61 in the Nainais-Kohero Pegmatite Belt, as well as a large number of pegmatites in the Karibib Pegmatite District on Okatjimukuju 55 and Etusis 75. A pegmatite on Etusis 75 is also exceptional due to its diversity of aluminium phosphates.

Tourmaline and beryl of gem quality can be found in many pegmatites of the Karibib Pegmatite District. The most important sources of gem tourmaline are the pegmatites of the Usakos Tourmaline Mine, Neu Schwaben 73, Otjua 37, Albrechtshöhe 44, Otjimbingwe 104, Omapyu 76 and Otjakatjongo 4.

Heliodor from the Hoffnungsstrahl pegmatite, Rössing Mountain, was one of the first gem minerals of spectacular quality discovered in Namibia (Kaiser 1912). Associated with common beryl, precious aquamarine has been recovered from a number of pegmatites, which belong to the Karibib Pegmatite District, like Donkerhuk 91, Tsaobismund 85, Salem 102 and Wilsonfontein 110. Morganite, the rose coloured variety of beryl, occurs at Etiro 50, and together with aquamarine at Rubicon. Emerald of bright yellowish green colour but low quality is found in a pegmatite on Neuhof 100 in the Maltahöhe District (Schneider & Seeger 1992). The rare beryllium mineral milarite has been found in many well-shaped, hexagonal crystals of bright yellow colour at Jasper's Pegmatite, Rössing Mountain.

In some pegmatites, the quartz of the core zone can be either of high purity or consisting partly of rose quartz. Whereas rose quartz is common in poorly fractionated pegmatites, like Tsaobismund 85, deeply coloured high quality is scarce and recovered from only a few places, e.g. Border

155 and Mickberg 262 in the Karasburg District, or from the Roselis Mine near Rössing Mountain area (Schneider & Seeger 1992).

Miarolitic cavities occur in granites of the Mesozoic anorogenic complexes. Best known are the topaz, aquamarine, heliodor and smokey quartz occurrences of Klein Spitzkuppe and nicely twinned crystals of bertrandite are also found here. Other topaz occurrences are Etemba 135 and Anibib 136. Topaz found in pegmatites is usually blue in colour, but not euhedral and scarcely of gem quality, for example at Etiro 50, Rubicon and Okatjimukuju. Peralkaline pegmatites of the Amis Valley, Brandberg, carry a variety of rare earth minerals and large astrophyllite clusters up to 3 cm in size.

The uranium mineralisation in alaskitic granites at Rössing is of great economic importance. Rössing Uranium Mine has created Namibia's largest open pit, but only a few remarkable minerals have been found, like boltwoodite and betafite.

Complexes of alkaline rocks and related carbonatites host quite a number of interesting minerals including different species of the pyrochlore group, monazite, xenotime, bastnaesite, synchisite, zircon, fergusonite and others. At Eureka 99, monazite in crystals up to 15 cm in size have been found (Steven 1993). Due to a very low radioactivity, the honey-brown mineral is not metamict but translucent to clear, displaying its full cleavage system. The Okorusu Mine near Otjiwarongo is well known for euhedral fluorite in various colours, and carbonatite dykes near Swartbooisdrif on the Kunene host blue lapidary grade sodalite (Menge 1986).

Good mineral specimens in Karoo basalts occur in Namibia and zeolites, calcite and quartz are the most common minerals known from the Etendeka and the Kalkrand Formations. Optical calcite was once recovered near Mariental, and from the Goboboseb Mountains W of Brandberg, lustrous crystals of amethyst and smoky quartz are recovered, many of them with skeletal growth and abundant fluid inclusions. Prehnite and analcime come from the same area.

The phonolite intrusions on Aris 28, south of Windhoek, are also recognised as a mineral locality. They contain numerous vesicles with a stunning abundance of micro-minerals (Garvie et al. 1999). Some very rare minerals have been found here, such as kanemite, villiaumite, makatite and tuperssuasiaite, the latter in crystals far superior to those from the type locality in Greenland.

The most notable mineral in metamorphic rocks is the orange spessartite garnet of the Hartmann Mountains in Kaokoland. These exceptional gems are amongst the most brilliant cut stones in the world, and are highly regarded. Light blue, translucent cordierite has been found at several places in garnet-cordierite schist of the Kuiseb Formation between Swakopmund and Henties Bay and also occurs, weathered from schists and gneisses, as water-worn pebbles in riverbeds (Schneider & Seeger 1992). Twinned staurolite occurs at Gorob in the Namib Naukluft Park and euhedral crystals of staurolite, mostly untwinned, can be collected from the banks of the Swakop River, immediately S of Okahandja. At Etemba 135, blue dumortierite is exposed in small narrow lenses, and although its origin is ambiguous, it is probably metamorphic.

The manganese ores of Otjosondu about 125 km NE of Okahandja, are associated with banded iron formation and have undergone Damaran metamorphism. The complex mineralogy includes jacobsite, braunite, hausmannite, bixbyite, hollandite, pyrolusite, psilomelane, calderite, spessartite, celsian, rhodonite and Mn-aegirine-augite (Schneider 1992).

A large variety of minerals is known from the old Arandis Tin (or Stiepelmann) Mine, some 25 km N of Trekkopjie. These include axinite, danburite, scapolite, vesuvianite and shiny brown tourmaline (dravite?). Two proposed new minerals, first described from Arandis, are however not valid species. Arandisite is a mixture of quartz and cassiterite, and stiepelmannite is identical to florencite-(Ce). Many specimens of epidote and scapolite are known from the Rössing Mountain and the Khan River Valley.

The world's largest alluvial diamond deposit, which has produced over 70 million carats since 1908 occurs along the Namibian coast. 95 % of all stones are of gem quality, they have good crystal shapes with curved octahedral and/or dodecahedral faces. The largest diamond weighed 246 carats (Schneider & Miller 1992).

Even bird and bat guano have yielded some extraordinary minerals. Stercorite from Ichaboe Island near Lüderitz described in 1850 was in fact the first new mineral ever reported from Namibia, and some unusual minerals, such as swaknoite, nitromagnesite, whitelockite, biphospammite, brushite, dittmarite, hannayite, struvite, newberyite and mundrabillaite have been identified from bat guano in the Arnhem and other caves (Martini 1992, Marais et al. 1996).

# 5. Palaeontology

by Martin Pickford and Brigitte Senut

Namibia's onshore fossil record spans three periods of the Earth's history, the Neoproterozoic, the Late Palaeozoic to Mesozoic and the Cenozoic. Proterozoic fossils occur in Neoproterozoic rocks of the Otavi and Nama Groups of the Damara Sequence. Various basins of the Karoo Sequence contain marine and terrestrial fossils of Permian to Triassic age, while Cretaceous fossils are found in isolated outcrops at Gross Brukkaros and in the S of the country. Early Miocene to recent fluvial, aeolian and littoral marine deposits of southwestern Namibia have yielded a wealth of fossils, and karst features in the carbonates of the Otavi Group contain Middle Miocene to recent fossils. Palaeontological studies have contributed greatly to the understanding of Namibia's geology, and a number of Namibian sites are ranking as world-class sites.

The oldest fossils so far recognized in Namibia are stromatolites and oncolites with an age of approximately 830 to 760 Ma (Pirajno et al. 1993). They occur in over 5 000 m thick dolomites of the Otavi Group of the Damara Sequence, which accumulated in a shallow shelf environment. There is a high diversity of types, with stromatolites typical of shallow water with high photic levels, as well as oolites which grew in deeper water with lower light levels. It seems that reef-type stromatolites formed an incomplete barrier between the open sea to the S and quiet lagoons on the landward side to the north. Some of these lagoons were extremely extensive and produced individual beds containing stromatolites that have been mapped for 160 km along strike and 10 km along dip. The cyanobacteria, which produced the stromatolites and oolites were the dominant life forms during most of the Proterozoic, and were largely responsible for the production and maintenance of the Earth's oxygen-rich atmosphere from about 2000 Ma onwards. The Otavi stromatolites therefore played a major role in the evolution of the Earth's atmosphere towards the end of the Proterozoic, since they were so wide-spread and existed over such a long time span.

The Vendozoan fossils of the Nama Group are amongst the best preserved and most extensive of their kind in the World. They provide information essential for understanding the palaeo-biosphere in the Neoproterozoic leading up to the so-called "Cambrian explosion" of different life forms some 550 Ma ago. The Nama Group spans the end of the Proterozoic to the onset of the Cambrian (~ 560 to 530 Ma) (Runnegar & Fedonkin 1992) and many genera and species of Vendozoans are based on Namibian material. The abundant Vendozoan fossils contained in the Nama Group are an enigmatic group of organisms about which there is much debate. Most of them were soft-bodied creatures such as *Pteridinium*, *Rangea* and *Ernietta*, which lived on or in seafloor sediments in shallow water (Gürich 1933), but other's, like for example *Cloudina*, possessed calcareous shells (Germs 1972). Fossils of the Nama Group illustrate the evolution of the Metazoa, and represent one of the few rock records, which span the period during which macroscopic, muscle-powered organisms evolved. The Nama Group has yielded the earliest evidence of the development of body protection in organisms. Presumably, these body coverings acted as protection against the muscle-powered metazoan predators, which left abundant traces of their presence in the sediments.

Sediments of the Karoo Sequence have accumulated in intra-cratonic settings and crop out extensively in Namibia. Karoo deposits cover a considerable geological time span from the Late Carboniferous to the Early Cretaceous, and their Mesozoic part has yielded important fossils ranging in age from 260 to 200 Ma. Several taxa of mammal-like reptiles and plants have been created for Namibian Karoo fossils. At the time of Karoo sedimentation, Africa, Antarctica and South America were part of a single landmass, Gondwanaland.

In Namibia, the basal Late Carboniferous to Early Permian Dwyka Formation consists predominantly of diamictites of glacial origin interbedded with shales, and is therefore poorly fossiliferous. However, fossils so far recorded include miospores, radiolaria, foraminifera, annelida, conulariida, mollusca, brachiopoda, bryozoa, porifera, echinodermata, fish and trace fossils.

The succeeding Permian Ecca Group has yielded abundant fossils from numerous localities. Perhaps the most famous deposit occurs within the Whitehill Formation, which contains complete skeletons of the sea-going reptile *Mesosaurus*, along with plant remains and intvertebrates. The

"Mesasaurus Sea" comprised an inland water body that extended through Brazil, Namibia, South Africa and parts of Antarctica, and is named after the aquatic reptile found in its sediments. This reptile also provided the first palaeontological evidence supporting Wegener's theory of continental drift, which was also postulated by du Toit (1927). *Mesosaurus* has a restricted distribution in southwestern Africa and Brazil which indicates that the two areas had once been much closer together than they are today. Further studies have amply confirmed the overall similarities in depositional successions and fossil content in Brazil and Namibia during Karoo times. There is also an abundance of fossil wood, especially in northwestern Namibia, where fossils forests occur (see 2.26). The Namibian Ecca deposits have also yielded stromatolites, miospores, crustaceans, mollusks, fish, amphibians and trace fossils.

The Triassic Omingonde Formation of the Karoo Sequence is of particular interest because it has yielded an abundance of amphibians, mammal-like reptiles and a thecodont. Several of the specimens from the Omingonde Formation consist of articulated skeletons. The overlying Etjo Sandstone of Late Triassic age is well known for the dinosaur tracks at Otjihaenamaparero (see 2.25), which have been developed into a tourist attraction. The herbivorous dinosaur *Massospondylus* was discovered in the Etjo Sandstone at Waterberg in 1997.

Cretaceous sediments are extremely rare in Namibia, which is why the fossil-bearing crater-lake sediments of the Cretaceous Gross Brukkaros (see 2.2) are of great importance. Among the fossil material recovered are pieces of wood, plant stems, seeds, leaves and flowers. Specimens of *Brachyphyllum* are relatively common, other identified remains include *Equisites*, *Pseridophytes*, an angiosperm flower with five main petals and seeds of *Carpolithus*. The fossils of Brukkaros together with other occurrences of the same age in South Africa and Botswana suggest that the Cretaceous vegetation of the subcontinent comprised Podocarp-Araucaria coniferous forest.

Marine Cretaceous sediments in Namibia are restricted to one fossiliferous deposit near Bogenfels, where gastropods, oyster shells and some ammonites of Cenomanian age were recovered (Klinger 1977). Shallow marine sediments of Early Eocene Age with a rich and highly diverse fauna including corals, bivalves, gastropods and a high diversity of sharks occur at Langental in the "Sperrgebiet".

Cenozoic deposits in the Sperrgebiet have yielded well over 100 animal species of which more than 80 are based on Namibian fossils. These fossils are contributing to the bio-stratigraphy of marine terraces and hence are important in the exploration for diamondiferous marine sediments, since diamonds only occur in sediments of a certain age. Researchers have recognized three major faunal assemblages of molluscan chronofaunas. With the additional help of mammalian bio-stratigraphy from the terrestrial facies, ages of Late Miocene, Late Pliocene and Late Pleistocene to Holocene respectively could be attributed to the three assemblages. Common fossils are various species of the mollusc *Donax* and the oyster *Crassostrea margaritacea*. The youngest assemblage contains records of the extinct equid *Equus capensis* as well as Middle Stone Age artifacts.

Many of the fluvial sediments in the "Sperrgebiet" also contain abundant fossils. Early Miocene ostracods, molluscs, frogs and mammals such as *Myohyrax* and *Apodecter stromeri* occur at Elisabeth Bay, while likewise Early Miocene rodents, macroscelidids, ruminants, suids, carnivores and rhinocerotids have been found at Fiskus. The most important fossil occurrence in the northern Sperrgebiet, however, is by far Elisabethfeld. It has yielded many thousands of fossil specimens representing a highly diverse fauna of Early Miocene age. Other occurrences are Grillental, Langental and Glastal.

Early to Middle Miocene fossils occur in abundance in old terraces of the Orange River. Early Miocene petrified tree trunks as well as small and large mammals and reptile remains have been found at the Auchas Diamond Mine. The famous Arrisdrift site upstream from Auchas has yielded more than 10 000 Middle Miocene fossils belonging to over 30 mammalian taxa, many of which are new to science. In addition to mammals there are also crocodiles, tortoises and other reptiles, birds and fishes, as well as a few plants. The local environment must have been considerably more humid than it is at present, and the climate was probably subtropical rather than temperate.

The Tsondab Sandstone Formation of the Namib Desert also contains abundant fossils, the commonest of which are rhizoliths and termite bioconstructions. The snail *Trigonephrus* and vertebrates, such as burrowing mammals have been found widespread. The most interesting fossil from these aeolianites, however, are the eggshells of struthian birds. The macrostructure of these eggshells changed rapidly during the Neogene, and they are therefore valuable tools for bio-stratigraphic correlation.

Middle Miocene to recent karst breccias have developed in Proterozoic carbonates in the Otavi Mountainland and in Kaokoland, and these are extremely rich in fossils. The Karstveld of northern Namibia represents a major palaeontological resource, one of the most comprehensive in Africa for the period spanning the Middle Miocene to recent. Investigations and research has led to the discovery and identification of hundreds of thousands of fossils. The Berg Aukas palaeo-cave has yielded by far the most extensive series of micro-mammal faunas known in Africa. It made world headlines in 1991 when the first known Miocene hominoid south of the Equator, *Otavipithecus namibiensis*, was discovered. However, the bulk of the mammals from Berg Aukas consists of rodents and bats, frogs and lizards, and the breccias yield few large mammals, and of the ones that have been collected, primates predominate. Other palaeocave deposits in the Otavi Mountainland are Harasib, Jägersquelle, Nosib, Aigamas, Rietfontein, Friesenberg, Gabus and Asis/Kombat. Elsewhere in Kaokoland, Plio- to Pleistocene karst deposits have yielded small- and medium-sized mammals and gastropods.

Pleistocene fluvial deposits in Namibia are generally poorly fossiliferous, but have yielded some interesting material, such as the calotte of the Orange River Man, an archaic *Homo sapiens*, and the Otjiseva human skeleton. The country is also endowed with abundant lithic cultures, ranging in age from Oldowan (Early Pleistocene), through Archeulean, Middle Stone Age, Late Stone Age to Iron Age.

# 6. Hydrogeology

by Jürgen Kirchner

Namibia with its position on the southwestern coast of the African continent has an arid climate similar to the west coasts of South America and Australia. Namibia is the driest country S of the Sahara with no permanent rivers except for the border rivers Kunene, Kavango and Zambesi in the N and the Orange River in the S, all of which have their sources outside Namibia.

Most of Namibia falls within the summer rainfall region with the main precipitation in the months of January to March. The average precipitation varies between less than 50 mm per year in the southwestern parts and 700 mm per year in the eastern Caprivi, while the mean gross evaporation ranges between 2500 and 3800 mm per year. NE of a line from Ruacana via Windhoek to Gobabis rainfall variability is below 30 %, whereas towards the coast it rises to about 90 %.

Given these circumstances it is not surprising that most of the country depends heavily on groundwater resources. More astonishing, however, is that there are so many springs, although many of them are low yielding and some not even permanent. The term "fontein" or "quelle" as part of many of the farm and place names is evidence of this fact. Furthermore, "Ai" for water, as in Aigams, the Nama name for Windhoek; Ai-Ais or other names like "Aub" indicate water in many of the original place names. In fact all old settlements were at springs or where river alluvium provided sufficient water. Old maps show most of these watering points, like for example, the "Kriegskarte" which was printed in Berlin in 1904.

Numerous small farm dams augment the boreholes, at least for part of the year, and assist in groundwater recharge. Only the major centers receive a greater proportion of their water demand from a few large dams, while larger parts of Ovamboland depend on water abstraction from the Kunene River. It must not go unmentioned that Windhoek's water reclamation plant contributes substantial amounts to the city's demands. The plant was recently enlarged to a capacity of about 7 million $m^3$ per annum, while the city's consumption is in the order of 18 million $m^3$ per annum.

There are an estimated 40000 producing boreholes in the country. They vary greatly in depth, water level and yield, but the average depth is 76.5 m, with the water level rising to 32.4 m and the average yield is 5.8 $m^3$ per hour. Three out of four recorded boreholes, including the dry ones, yield more than 1.1 $m^3$ per hour. Water level and abstraction data are recorded monthly for the more than 100 water supply schemes.

Grootfontein, Tsumeb, Outjo, Omaruru and Windhoek operate their own water schemes, with Windhoek also being supplied by NamWater from the Von Bach and Swakoppoort Dams. In the rest of the country NamWater is responsible for the bulk water supply to schemes varying in size from schools in rural areas to larger areas, like the central Namib area, that include a number of large consumers, such as the municipalities of Swakopmund and Walvis Bay, and the Rössing Mine.

The main constituents for the classification of groundwater quality are total dissolved solids, sulphate, nitrate and fluoride. There are a few areas in Namibia, where groundwater is unsuitable for human consumption and stock watering, mainly the "Saltblock" W of Mata Mata on the Auob River at the South African border and an area E of Kalkrand. Other occurrences of unsuitable groundwater are more localized. An average value of 1249 mg/l of total dissolved solids, 209 mg/l sulphate, 12.3 mg/l nitrate and 1.1 mg/l fluorine has been measured. Most natural groundwater occurrences have a pH of slightly more than 7.0 and are hard (Kok 1964, Department of Water Affairs 1978).

Borehole observations have shown that water levels tend to drop in normal and below average rainfall years due to aquifer drainage and abstraction. The rainy season itself may be indicated by a reduced fall in water levels or even a slight rise. During very high rainfall years with an annual precipitation of more than twice the average, the aquifers are restored to about maximum level. Those parts of an aquifer furthest away from the point of recharge recover later than parts close to the point of recharge. Water levels near the recharge point therefore drop at a slightly increased rate while the lowermost parts of the aquifer can still be recharged. Only if another very high rainfall year follows within the next two years, aquifers are fully recharged to their maximum capacity. Past experience has shown that very high rainfall years occur about once in thirty years. Furthermore, the tendency for decreasing rainfall during the past two decades may lead to longer intervals between complete recharge events.

During recent years first attempts have been made to artificially recharge groundwater. The Omdel Dam upstream of Henties Bay has been built to infiltrate runoff of the Omaruru River and store it in the Omaruru Delta aquifer. Evaluation of the recharged quantities is still under way. Infiltration tests have also been carried out successfully on the Windhoek aquifer and an infrastructure is created to infiltrate purified dam water on a routine basis. Because of the high evaporation losses from dams, other potential aquifers such as the alluvial Osona Aquifer at Okahandja and the Oanob Aquifer at Rehoboth are also being considered for potential artificial recharge.

Two other techniques to enhance natural recharge have been developed and applied in the past, the "Grundschwelle" and the sand storage dam. The "Grundschwelle" is a stone, brick- or concrete wall like a weir that suppresses subsurface runoff in the alluvial material of a river and promotes recharge through fractures in the underlying bedrock. A sand storage dam is a shallow wall in rivers carrying arenaceous sediment. The shallow wall, about 0.5 to 1 m above the level of the alluvial material, has a v-notch that reaches down to the alluvial surface. During runoff events coarse material is deposited while water with the finer sediment can drain through the notch when the flood recedes. After the flood the dam wall is raised by about 0.5 to 1 m and the process can be repeated until the dam has reached the optimum height. Water is stored in the retained coarse sediment and evaporation is substantially reduced. A number of these dams can be found in the vicinity of Windhoek, for instance at Ovitoto, Otjimbingwe, and on Neudamm and Ondekaremba (Stengel 1968, Hellwig 1978).

Unfortunately, a large part of Namibia is underlain by metamorphosed Precambrian rock units. This means, in terms of hydro-geological properties that the primary porosity tends to be low and higher permeabilities are restricted to secondary jointing and fracturing or carbonate dissolution.

There are three artesian areas in the country, the Stampriet Artesian Basin around the lower Auob and Nossob Rivers E of a line from Kalkrand to Asab; the Maltahöhe Artesian Area E of the Zaris Mountains; and an area NW of Tsumeb.

The Stampriet Artesian Basin is by far the largest and most important aquifer. It stretches roughly from a line from Uhlenhorst to the confluence of the White and Black Nossob Rivers and further E beyond the Botswana border in the N; the watershed is between the Auob and Fish Rivers on the edge of the Weissrand Plateau in the W and from Tses eastwards beyond the Botswana border in the S. Some 6000 boreholes supply water for stock watering and limited irrigation mainly S of Stampriet and around Leonardville.

The confined aquifers consist of a lowermost Nossob Sandstone separated by black shale from the overlying Auob Sandstone which is again separated by blue shale from the upper unconfined Kalahari aquifer. The Karoo formations dip gently in a southeasterly direction while the Kalahari Group sediments increase in thickness from 0 to about 300 m from W to E. The groundwater is artesian and free flowing only along the upper Auob River valley down to between Stampriet and Gochas; and in the Nossob valley from upstream Leonardville to Aranos. Outside the valleys the potentiometric surface remains generally below ground level. Discharge rates of more than 100 $m^3$ per hour have been recorded from boreholes, although mean yields are about an order lower.

Sub-outcrops of the aquifer are recharged along the northern boundary of the area through the alluvium of the rivers originating in the mountains S and E of Windhoek, as well as S of Rehoboth. A second recharge area lies in the Weissrand Plateau and recharge also occurs from the water-bearing Kalahari layers.

$^{14}C$ analyses of water in the aquifer yielded apparent very old ages of about 15000 years in the area surrounding Stampriet, about 25000 near Aranos and 35000 S of the Nossob River near the Botswana border (Vogel & Van Urk 1975). In all three aquifers the water quality tends to deteriorate in a southeasterly direction, where total dissolved solids can exceed 20000 mg/l in the salt-block SE of Gochas (Tredoux & Kirchner 1980, 1981, 1985).

A smaller artesian area lies SW of Maltahöhe. Here, the fractured Schwarzkalk of the Nama Group is recharged in the Zaris Mountains from where it dips eastwards under Schwarzrand shale. However, of the boreholes drilled only a few have a high yield and are used for stock watering and small-scale irrigation.

Another artesian aquifer occurs in Kalahari beds N of the western Otavi Mountainland stretching from the Oshivelo area northwards and westwards into the Etosha National Park. It is assumed that the recharge area lies near the foothills of the Otavi Mountainland (Hoad 1993).

Within the Namib Desert floodwaters of the Koichab, Tsauchab and Tsondab Rivers end in pans or vleis and fail to reach the Atlantic Ocean. However, their groundwater flows westwards along buried river courses under the dunes and can be found at shallow depths on the beaches of the Atlantic coast at Anigab, Fischersbrunn and Conception Bay respectively. The course of the Kuiseb River is also threatened by the migrating dunes but so far major floods have managed to clean its bed. The Swakop and Omaruru Rivers, and the rivers further N continue to release their waters into the ocean during good rainy seasons.

The Koichab, Kuiseb and Omaruru rivers are important water supplies for the coastal towns. The Koichab Scheme supplies the town of Lüderitz from a well field upstream of Koichab Pan some 100 km inland at the foot of the Namib dunes. The annual abstraction is in the order of 1.5 million $m^3$. The Kuiseb River with its Swartbank and Rooibank Schemes produces around 6.5 million $m^3$ per annum and the Omaruru Delta with its Omdel Scheme about 6.4 million $m^3$ per year. These three schemes supply the towns of Walvis Bay, Swakopmund and Henties Bay, as well as the Rössing Uranium Mine and the nearby town of Arandis, 60 km inland from Swakopmund. Increasing demand and unsustainable abstraction rates from the Omaruru Delta have led to the construction of the Omdel Dam upstream of the well field to artificially recharge the Omdel aquifer (Stengel 1966, Tordiffe 1990).

Quartzite of the Auas Formation, which forms the Auas Mountains S of Windhoek, dips at shallow angles northwards under mica schist of the Kuiseb Formation. The fractured and faulted quartzite is recharged in these mountains and also from the upper parts of the Usib River catchment S of the mountains. Groundwater circulates at great depth and used to reach the surface as hot springs just E of the Windhoek Municipality building at recorded temperatures of 78 °C, and a number of springs range in temperatures from 40 °C to 60 °C at the Wasserberg in Klein Windhoek. The main water-bearing structures are near vertical breccia-filled faults that strike roughly N-S. The aquifer has been developed to meet increased demands and today the well field comprises some 40 producing boreholes S, SW and SE of the town. The mean annual abstraction is about 2 million $m^3$, and because of this, the water level in the former springs has dropped 15 m or more below ground level. Apparent $^{14}C$ ages between 6 000 years on the northern slopes of the Auas Mountains and 20 000 years at the Pahl Spring indicate a velocity of groundwater flow of about 1 m per day.

The comparatively large storage capacity of the Windhoek aquifer, which is estimated at between 10 and 25 million $m^3$, and the high evaporation rate from dams in Namibia recently led to the Windhoek Municipality conducting artificial recharge trials with purified water from the von Bach Dam. First results were promising and may lead to better usage of the scarce water resources available (Ministry of Agriculture, Water and Rural Development 1993).

The Otavi Mountainland is at the watershed between the Okavango Basin in Botswana, the Etosha Pan and the Huab River catchment. It is a synclinorium that consists largely of carbonate rocks of the Otavi Group of the Damara Sequence, and calcrete covers much of the ground. Although rainfall is comparatively high at 550 mm per annum, no drainage system has developed due to the high permeability of the surface layers. A number of Namibia's highest yielding springs are located in the area indicating large groundwater reserves. Further indications of large groundwater reserves were also given by the high pumping rates of mines such as Tsumeb, Kombat, Abenab and Berg Aukas. These mines have abstracted collectively up to about 20 million $m^3$ per annum without major effects on water levels in the wider surrounding area.

The southernmost syncline has been developed in the 1980s for abstraction into the Grootfontein-Omatako Canal. Kombat Mine is already discharging into the canal, and the former Berg Aukas Mine has recently been connected. Investigations are under way to link the former Abenab and Tsumeb Mines to the canal (Seeger 1990).

# 7. Mining

## 7.1 A historical perspective

The earliest mining activities in Namibia date back some 400 years according to archaeological evidence of copper smelting in the area W of Windhoek. It is also well established that the indigenous people of the Otavi Mountainland made use of the abundant copper ores in the area for many centuries. The first written accounts of Namibia's mineral resources can be found in the reports of early explorers, who came to the country in the second half of the 18$^{th}$ century. The discovery of guano on Itchaboe Island in 1828 greatly stimulated interest in the area, and eventually led to the discovery of the Pomona copper-silver lead occurrence in 1857. Just a year earlier, 1856, the famous explorer Charles John Anderson had opened up Namibia's oldest mine, the Matchless copper mine. Intensive exploration for copper was carried out in the country around that time resulting from the copper rush that followed the 1853 discovery of copper in the Cape Colony to the S. This exploration led to some production at the Natas, Pot and Sinclair Mines.

After Namibia had become a German colony, the "Deutsche Kolonialgesellschaft für Südwestafrika" was founded in 1885 with the aim to explore for minerals and the first mining ordinance was promulgated in 1888. The limited success of the "Deutsche Kolonialgesellschaft für Südwestafrika" did not discourage other companies, a number of which explored the country from the Kaokoveld in the N to the Richtersveld in the S. Many interesting discoveries were made, some of which developed into mining ventures around the turn of the century.

Mining at the famous Tsumeb copper deposit started in 1906, and the first smelter was established there in 1907. Other deposits in the Otavi Mountainland were also exploited and a railway line was built for ore transport to Swakopmund. The holding company, the "Otavi Minen- und Eisenbahngesellschaft", developed into one of the most profitable ventures in Namibian history. The Khan copper mine E of Swakopmund came on stream in 1905. Copper was also produced from 1907 onwards at Otjizonjati, E of Windhoek, and at Gorob and Hope SE of Swakopmund. The first tin discoveries were made in 1908 near Omaruru, and by 1912, 14 tin mines had been established.

The most important discovery which impacted Namibia's recent history, was without doubt made in 1908, when Zacharias Lewala found the first diamond near Lüderitz. Within a few months, the entire coastal strip S of Lüderitz was pegged, and enormous investments and efforts were made. A 30-km long narrow gauge railway line was built in 1909 from Lüderitz to the S, on which ore was pulled by mules, before locomotives were introduced in 1912. A pipeline and a pump station were erected to supply water to the claims in the desert, and the early hand-operated sieves and crushers were later replaced by more mechanised equipment. These developments meant that by 1913, 20 % of the World's diamond production came from Namibia.

In addition to diamonds, copper and tin, quarrying of marble started in the Karibib area in 1909, and a limited amount of semi-precious stones such as topaz, aquamarine, heliodore and tourmaline were recovered mainly from pegmatites in the same area. Gold, tungsten and manganese were also discovered at this time, but no production occurred.

Mining virtually came to a standstill in 1914 with the outbreak of the First World War. However, South African Union troops soon took over the country and diamond production resumed in 1915. After the War, the "Otavi Minen- und Eisenbahngesellschaft" resumed full production at Tsumeb in 1921, and in 1922 the first germanium ore was recovered as a by-product. In 1927, a first shaft at the Klein Aub copper mine SW of Rehoboth was opened up, and the latter part of the decade saw development of the Abenab, Baltika and Berg Aukas vanadium mines. The gold deposits of Ondundu N of Omaruru were discovered in 1917, and production started in 1922. Tin production resumed and in particular the Uis mine was developed. The Krantzberg and Natas mines began to produce tungsten ore in 1933, and in the same year gold mining started in the Rehoboth area.

After the First World War, diamond production had been apportioned between a number of companies, and in 1920, Sir Ernest Oppenheimer amalgamated nearly all these companies into the "Consolidated Diamond Mines of South West Africa" (CDM). CDM was soon to become one of the most successful diamond mining companies in the world. A large recovery plant was established in 1924 at Elizabeth Bay, and in 1928, the world's richest occurrence of gem quality diamonds was discovered N of the mouth of the Orange River. As a result of the growing importance of the southern deposits, CDM headquarters moved to Oranjemund in 1943. Meanwhile, diamond mining also occurred in the area N of Lüderitz, and small settlements were established in the middle of the desert at Charlottenfelder, Holsatia and Fischersbrunn.

After the Second World War, all properties of the "Otavi Minen- und Eisenbahngesellschaft" were put on sale by the Custodian of Enemy Property. A syndicate of international companies bought the assets, and Tsumeb Corporation was established. The Kombat mine started production in 1962, and in 1970 the Otjihase mine was developed. Rosh Pinah, one of the largest zinc deposits in southern Africa, was discovered in 1964 and at this time the Uis tin mine developed into the largest hard rock tin mine in the World. With the Otjosondu manganese mine and the Okorusu fluorspar mine, two other big producers came on stream, increasing the importance of mining in Namibia.

The most important developments of the more recent past are the opening of the Rössing uranium mine in 1976, the development of mining methods to recover diamonds from the sea floor since the late 1960s, and the discovery of the Navachab gold deposit in 1985. From modest beginnings, the mining industry of Namibia has developed into the backbone of the country's economy, and with promising potential for the discovery of new mineral deposits, coupled with a keen interest shown by international exploration companies, Namibia's mining industry should prosper well in the future.

## 7.2 Present mineral production and exploration

Mining plays a vital role in the Namibian economy, and currently contributes about 12 % to the Gross Domestic Product. Export earnings amount to 36 % of total Namibian export income and mining is paying over 700 million N$ in taxes annually. Mining, although not labour intensive, is also an important employer in Namibia. The mineral production of Namibia from 1990 to 2000 is given in Tab. 7.2.1.

**Tab. 7.2.1:** Mineral production in Namibia 1990–2000 (Source: Directorate Mining, Ministry of Mines and Energy).

| | 1990 | 1991 | 1992 | 1993 | 1994 | 1995 | 1996 | 1997 | 1998 | 1999 | 2000 |
|---|---|---|---|---|---|---|---|---|---|---|---|
| Diamonds [ct] | 673837 | 1 186874 | 1 549260 | 1 141352 | 1 312348 | 1 381757 | 1 402129 | 1 416334 | 1 465959 | 1 632860 | 1 541747 |
| Uranium [st] | 3 787 | 3 185 | 2 190 | 1 968 | 1 896 | 2 007 | 2 892 | 3 425 | 3 278 | 3 171 | 3 201 |
| Copper [t] | 25 596 | 31 327 | 31 285 | 29 308 | 27 373 | 22 530 | 14 904 | 17 879 | 6 500 | 0 | 5 070 |
| Lead [t] | 18 527 | 15 176 | 14 942 | 10 907 | 13 142 | 16 084 | 15 349 | 13 577 | 13 303 | 9 879 | 12 115 |
| Zinc [t] | 37 690 | 35 420 | 35 657 | 17 970 | 33 575 | 30 209 | 34 377 | 39 658 | 42 142 | 37 429 | 40 266 |
| Pyrite [t] | 138 925 | 126 119 | 164 190 | 113 703 | 121 634 | 103 140 | 90 735 | 94 585 | 28 174 | 0 | 11 967 |
| Gold [kg] | 1 605 | 1 850 | 2 025 | 1 953 | 2 430 | 2 099 | 2 145 | 2 417 | 1 882 | 2 008 | 2 400 |
| Silver [t] | 92 | 91 | 89 | 72 | 62 | 66 | 42 | 41 | 16 | 0 | 17 |
| Salt [t] | 157 224 | 141 368 | 120 835 | 132 585 | 356 865 | 429 779 | 355 868 | 492 780 | 536 180 | 503 479 | 542 948 |
| Fluorite [t] | 27 107 | 27 816 | 37 680 | 43 466 | 52 226 | 33 559 | 31 457 | 23 160 | 40 685 | 57 599 | 66 129 |
| Wollastonite [t] | 0 | 305 | 416 | 824 | 1 309 | 967 | 248 | 194 | 267 | 347 | 441 |
| Lithium minerals [t] | 1 268 | 1 192 | 1 162 | 793 | 1 362 | 2 611 | 1 971 | 632 | 247 | 0 | 0 |
| Marble [t] | 12 881 | 10 031 | 0 | 13 359 | 12 061 | 16 935 | 12 673 | 13 743 | 9 807 | 11 221 | 24 426 |
| Granite [t] | 5 437 | 7 890 | 7 313 | 2 952 | 11 585 | 4 518 | 5 218 | 5 675 | 6 676 | 7 222 | 5 866 |

ct = carat, st = short tons, t = tons, kg = kilograms

The diversity of Namibia's mines reflects the wide variety of metallogenic provinces. The diamond mines are by far the most important mining operations in the country. The precious stones have accumulated in alluvial deposits along the Namibian coast, both onshore and offshore, as well as in old terraces of the Orange River. So far, more than 70 million carats of diamonds have been recovered, 95 % of which are of gemstone quality. Mining of the beach deposits N of Oranjemund involves exposure of the diamondiferous gravel by overburden stripping using earth-moving machines, which include two large bucket-wheel excavators. The overburden sand is used to construct sea-walls which shift the beach up to 200 m seawards and thereby exposing gravels up to 20 m below sea level. The gravel is then removed by hydraulic excavators, and industrial-sized vacuum machines are used for the final "clean-up" process. After treatment in a recovery plant, which includes crushing, screening, dense media separation and X-ray sorting, a concentrate is sent to a central sort-house.

Diamonds are also recovered at Elisabeth Bay S of Lüderitz, where mining is carried out with mechanical excavators. Diamondiferous gravels of the Orange River terraces are recovered and treated at Auchas and Daberas, and the concentrates of both mines are forwarded to the central sort-house at Oranjemund for final treatment.

Near-shore diamond mining operations in shallow water are carried out by divers, who operate either from the beach or from small vessels, often converted fishing boats, which can be seen mooring at Lüderitz harbour. Gravel is sucked with large hosepipes on board the vessels or onto the beach, where a concentrate is recovered by small plants. Final treatment takes place in a central recovery plant in Lüderitz.

Pioneering work to recover diamonds form the seabed offshore Namibia has been carried out since the mid 1960s. Today, a fleet of 10 purpose-built vessels belonging to 3 companies, mine diamonds in water depths of up to 200 m, using drill bits and seabed crawlers. Some 50 % of Namibia's total diamond production currently comes from the ocean floor, a figure which is set to increase, as the onshore deposits mined since the early 20$^{th}$ century, become depleted.

Besides diamonds, Namibia is also one of the World's principal uranium producers. The Rössing Uranium Mine operates the World's second largest open pit uranium mine, which is located some 65 km inland from Swakopmund. The uranium mineralisation is hosted by alaskitic granites.

The production is monitored by a unique scanning system, which monitors the ore grade in loaded trucks with sufficient accuracy to establish cut-off grades. Mining is carried out at a rate of 43 000 tons per day with an average ore grade of 0,035 % uranium.

Base metal mining has a long tradition in Namibia, and to this day, copper remains an important commodity, followed by lead and zinc. Ongopolo Mining owns and operates several base metal mines and the smelter at Tsumeb. The Kombat Mine, some 51 km from Tsumeb produces ore grading 3,13 % copper, 0,93 % lead and 22 g/t silver. The Tsumeb-type Khuisib Springs Mine, located between Tsumeb and Grootfontein, was opened in 1996. At Otjihase, 20 km NE of Windhoek, cupriferous pyrite bodies are also mined by Ongopolo. Current ore production grades 2 % copper, 16 g/t silver, 0.35 g/t gold and 16 % sulfur as pyrite.

The Rosh Pinah Mine, located some 20 km N of the Orange River at the eastern boundary of Diamond Area No 1, is a major lead and zinc producer. The ore grades 7 % zinc and 2 % lead. A second zinc mine, the Scorpion Mine, is currently being developed just W of Rosh Pinah. These two mines, together with the zinc mines in northwestern South Africa, form one of the major zinc provinces in the World.

Despite numerous gold deposits occurring throughout Namibia, today's production originates from only one mine, the Navachab Gold Mine W of Karibib. Gold occurs in small quartz veins hosted by carbonate rocks, and the ore is mined in an open pit.

Industrial mineral production comprises a range of commodities, including fluorite, salt, wollastonite and lithium minerals. Namibia is one of Africa's largest salt producers, with two solar evaporation pan complexes at Swakopmund and Walvis Bay, the latter being responsible for the sharp increase in salt production in 1994, following the re-integration of Walvis Bay into Namibia. An acid-grade fluorite concentrate is produced at Okorusu Mine north of Otjiwarongo.

Namibia's dimension stone production comprises marble, granite, dolerite and sodalite, mostly derived from the central part of the country, except sodalite which is quarried in the far north near Swartbooisdrif. A cutting and polishing plant for dimension stone operates in Karibib. Semi-precious stone production takes place mainly on a small scale. The portfolio includes tourmaline, aquamarine, heliodore, morganite, garnet, topaz, amethyst, rose quartz and agate.

Exploration for the discovery of new mineral deposits continues throughout Namibia on a regular basis by both local and international companies. The most sought-after commodities are diamonds, base metals and gold, followed by industrial minerals and dimension stone. Diamond Area No. 1, which was an area off limits since the beginning of the 20th century, has recently been opened up for exploration and is attracting major international interest. Modern geophysical methods are applied to find occurrences covered by the young sediments of the Namib Desert and Kalahari Group. Diamond exploration on the sea floor also requires highly sophisticated methods. Hydrocarbon exploration is carried out in the Namibia offshore area, where a major gas field is also about to be developed. Over 600 million N$ were spent on exploration by private companies in Namibia between 1995 and 1999.

# 8. Geological attractions

The following part of the Roadside Geology Guide deals with individual geological attractions in Namibia. These are sites that are of such outstanding geological significance and interest, that a detailed, separate description is justified. In the excursion part (9.), reference is made to the following chapters.

## 8.1 Brandberg

by Axel Schmitt, Rolf Emmermann and John Kinahan

The Brandberg is one of the Cretaceous anorogenic complexes of northwestern Namibia, and from a spacecraft, it is one of the most eye-catching circular features visible on Earth. The isolated massif of granite, with approximate dimensions of 26 by 21 km, rises more than 2000 m above the surrounding peneplain of the Namib Desert. The edges of the roughly cupola-shaped plutonic body are dissected by numerous gorges, which penetrate the central part of the complex to form a radial drainage system. The summit, named Königstein, at 2573 m above sea level is the highest elevation in Namibia. The reddish colour of the weathered granite surfaces led to its German name "burning mountain" as well as its original Damara name "Daures" meaning a pile of ash. The Brandberg is a National Monument, and the main access is via the road from Uis leading to the "White Lady" rock painting in the Tsisab Gorge (Fig. 8.1.1).

The dominant plutonic rock type of the Brandberg is a medium grained biotite-hornblende granite showing little textural variation. Fine-grained enclaves near the contact are interpreted as a chilled margin facies. Segregations of a chemically evolved biotite-leucogranite locally form apophyses within the main granite. Numerous leucocratic dykes cut the granite massif and the adjacent country rocks. These dykes range in size from a few centimeters to about 3 m in width. Another plutonic variety is a coarse-grained pyroxene-bearing monzonite which is exposed in the western inte-

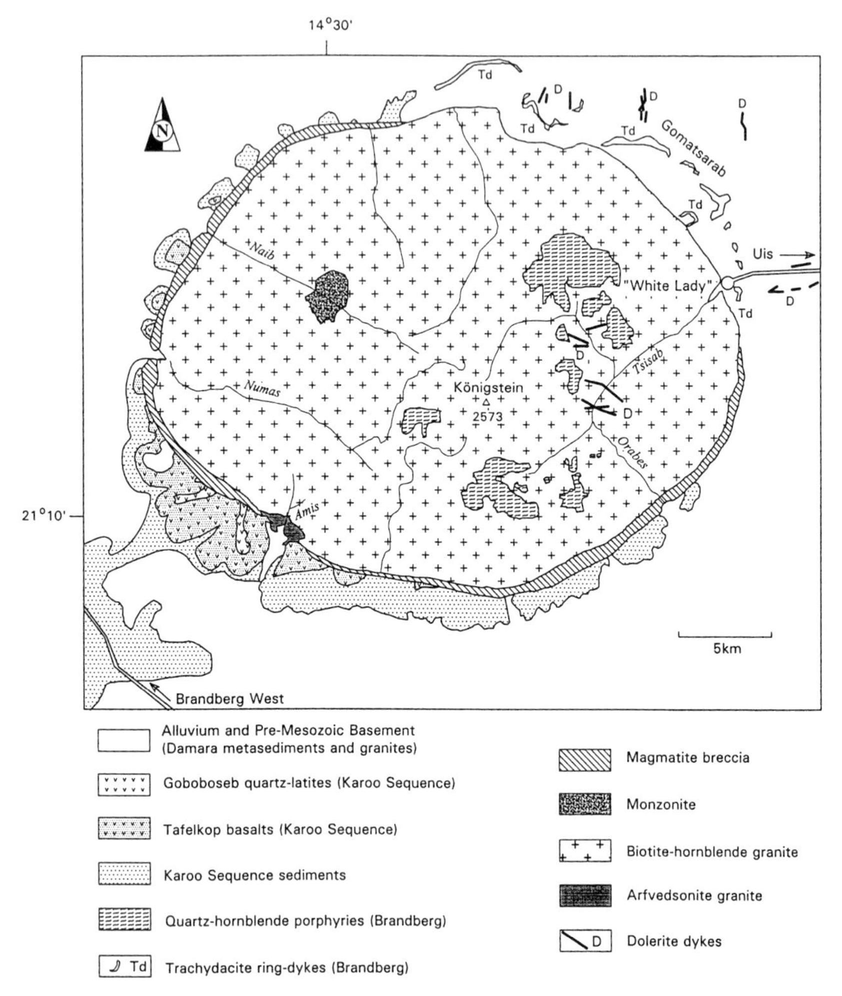


**Fig. 8.1.1:** The geology of the Brandberg.

rior of the massif. Contact relations suggest that the monzonite is older than the biotite-hornblende granite. Quartz-hornblende porphyries occur in the central south-eastern part of the mountain. The youngest intrusive phase is a small arfvedsonite granite sill, which crops out in the Amis Valley at the southwestern edge of Brandberg. This granite carries the amphibole arfvedsonite as the main dark mineral and is highly variable in mode and texture. An indurated hematised variety forms the prominent cliffs in the inner part of the Amis Valley. Related aplitic to pegmatitic dykes intruded the biotite-hornblende granite and the marginal Karoo volcanics in an area up to 1 km distance from the contact zone.

The dominantly medium-grained granite of the main intrusion is reddish on weathered surfaces but appears gray-greenish when fresh. Fine-grained rounded enclaves of similar composition range in size from a few cm to about 50 cm and are common in the marginal facies of the granite. Miarolitic vugs filled with coarse K-feldspar or quartz are occasionally present. In hand specimen, one can observe tabular plagioclase which are often rimmed by whitish K-feldspar. Biotite and hornblende are the most abundant dark minerals. Opaque minerals are titano-magnetite and ilmenite and accessory minerals include apatite, zircon, titanite, monazite and the rare titanite variety cheffkinite. The textures and compositions of the minerals indicate that the Brandberg granite crystallized from a relatively hot and dry magma. Differentiation of the magma produced late-stage leucogranites. These contain biotite as the almost only mafic mineral, although tourmaline is also occasionally present. Macroscopically visible granophyric intergrowth of quartz and orthoclase are common in the leucogranitic dykes, which indicates crystal growth during rapid cooling.

Radiometric dating of the main Brandberg granite has yielded an age of about 130 million years (Schmitt et al., in prep.). This immediately post-dates the peak volcanic activity in the Etendeka flood basalt province (see 8.6), and the surrounding Goboboseb Mountains, as well as the intrusion of the Messum Complex (see 8.18). Field evidence clearly indicates that the Brandberg post-dates at least the lowermost units of the Etendeka lavas.

A series of dark coloured trachydacitic ring-dykes is present to the north and northeast of the massif in the Gomatserab area. These dykes are sub-parallel to the rim of the intrusion and dip about 30° to 50° towards the center. In the uppermost reaches of the massif, few rafts of dark volcanic hornblende and quartz porphyries are preserved. These and the trachydacite ring-dykes can be related to the early volcanic phase of the Brandberg.

The surrounding country rocks of the Brandberg intrusion consist of Damaran granites and metasediments and overlying Karoo Sequence sediments and volcanics. The sediments comprise siltstones, altered to black hornfels in the contact aureole of the Brandberg, which are overlain by white, quartz-rich coarse conglomerates. They are in turn overlain by basalts capped by quartz latites. Remnants of the Karoo rocks are preserved in a collar along the western and southern margin of the massif and are down-faulted towards the contact. The angle of dip increases as the contact is approached and is near vertical at the contact, where clasts of country rocks occur within the granite forming a magmatic breccia.

A likely scenario for the genesis of the Brandberg is that heating and subsequent partial melting within the crust was triggered by emplacement of mantle-derived basaltic magma. Assimilation of crust-derived partial melts by the cooling and crystallizing basic magmas led to the formation of a hot and anhydrous hybrid granitic magma, which was capable of rising up to near-surface level.

On the way to the parking lot at the eastern edge of Brandberg, the trachydacite dykes can be seen to the N of the road, forming smooth dark hills weathered out of the older Damaran granites. Small roadside outcrops are present near the contact of the Brandberg massif. The trachydacite is a gray porphyritic rock with plagioclase phenocrysts up to 3 cm in length. The mafic mineral assemblage comprises pyroxene and hornblende, set in a fine-grained matrix of intergrown K-feldspar and quartz. Dolerite dyke swarms which cut the Damaran granites are also present close to the road. The dark, highly altered holocrystalline dolerites have a fibrous texture of plagioclase and contain relics of decomposed olivine phenocrysts. Similar dolerite dykes cutting the Brandberg granite have been described from the interior of the massif in the Tsisab Gorge and Bushman Valley (Diehl 1990).

The interior of the Tsisab Gorge is filled with debris, which carries huge granite boulders up to 3 m in diameter. The higher flanks of the gorge display well exposed surfaces of granite polished by erosion. At the lower flanks, multiple stages of onion-skin weathering typical for granites can be observed.

Along the southern margin of the Brandberg along the road to the defunct Brandberg West Mine, one can find highly altered pyrophyllite bearing Karoo siltstones within the Brandberg metamorphic aureole, which are locally quarried and manufactured into ornaments.

The archaeology of the Brandberg has been the subject of serious research for more than eighty years. Detailed surveys of the rock art have recorded more than 1000 sites, some with a hundred or more individual paintings. Although the most famous site, the Maack or "White Lady" Shelter, has given rise to several fanciful interpretations, systematic excavations in other parts of the mountain show that the area was inhabited by hunter-gatherer communities until the first appearance of nomadic livestock farming about 1000 years ago. Small bands of hunters evidently lived in the upper parts of the mountain during the dry season when little water or food could be obtained in the surrounding desert. The structural geology of the mountain, with its well-developed sheet joints, provides many small aquifers and where these emerge, rock painting sites are never far away.

In the rock art of Brandberg, human figures comprise more than 40 % of the images and among the many animal species depicted giraffe are often the most numerous. Few of the animals featured in the paintings are represented in bones recovered from archaeological excavations. Indeed, very few of the species in the paintings actually occur on the mountain itself which is far too rugged for most of them to ascend. This and other evidence, such as artifacts of crystalline quartz, marine shells and some metal objects, suggests that the people who inhabited the Brandberg also inhabited a far wider area. A clearer pattern of movement arose with the development of pastoralism when stock camps were established at remote waterholes and the herds were pastured far into the Namib Desert after the summer rain. In the dry season, however, pastoral communities would retreat to the upper Brandberg with its reliable waterholes and nutritious pastures, usually camping in the same places as their hunter-gatherer predecessors.

## 8.2 Brukkaros

by Volker Lorenz

Gross Brukkaros is an impressive inselberg in southern Namibia, located about 40 km west of the village of Tses. The mountain has a basal diameter of 7 km and a height of 600 m above the surrounding plain, which is itself at an elevation of about 1000 m above sea level. Brukkaros, because of its unique shape, has attracted general and scientific interest since the beginning of the century.

From a morphological point of view, Brukkaros forms a rather sharp and steep-sided ring-shaped ridge 3 km in diameter. This ridge surrounds a central basin, and therefore, the mountain resembles a volcano with a crater rim and a crater. Geological studies, however, reveal that Brukkaros is the erosional remnant of a completely different volcanic structure, and that it was originally surrounded by about 70 small maar volcanoes. Maar volcanoes are low-relief, broad volcanic craters formed by multiple shallow explosive eruptions.

Today, Brukkaros rises on top of a large uplifted circular area where reddish-brown, hard sandstones and shales of the Fish River Subgroup, Nama Group, are exposed (Fig. 8.2.1). They first dip away from Brukkaros and closer to it dip towards the mountain. These sediments were overlain in the Permo-Carboniferous by tillites and shales of the Dwyka Group of the Karoo Sequence of the great Gondwana glacial period. Remains of the Dwyka beds, which escaped erosion occur locally on the eastern and southwestern slope of Brukkaros. Overlying the shales and the locally preserved Dwyka beds and forming the steep upper cliffs are the Brukkaros beds, a series of sediments of fluvial and lacustrine origin. These sediments yield plant fossils, which assign them a Cretaceous age (Kelber et al. 1993).

Surrounding Brukkaros on all sides are more than 100 distinctly yellowish carbonatite dykes and 74 carbonatite pipes. Carbonatites are carbonate rocks of magmatic origin, often associated with kimberlites. The dykes represent fissures, which were filled by rising and finally solidifying carbonatite magma. The dykes are usually 30 cm to 1 m thick and can be traced for several hundred metres. The dykes are intersected by carbonatite pipes, which were the feeder pipes of the carbonatite maar volcanoes. The larger ones form little knolls or even hills and may reach a diameter of several tens of meters, or even more than 100 meters in exceptional cases (Kurzlaukis & Lorenz 1997).

Recent studies of the area surrounding Brukkaros, the Brukkaros beds and the carbonatite dykes and pipes, reveal an interesting volcanic history of Brukkaros and its surrounding volcanic field, which also sheds light on the geological history of the whole region. In Jurassic times, about 183 Ma ago, the Dwyka and overlying Ecca and Whitehill shales (about 300 to 280 Ma old) were regionally intruded by basaltic magma in the form of one or two thick, flat-lying dolerite sheets. On top of these dolerite sheets,

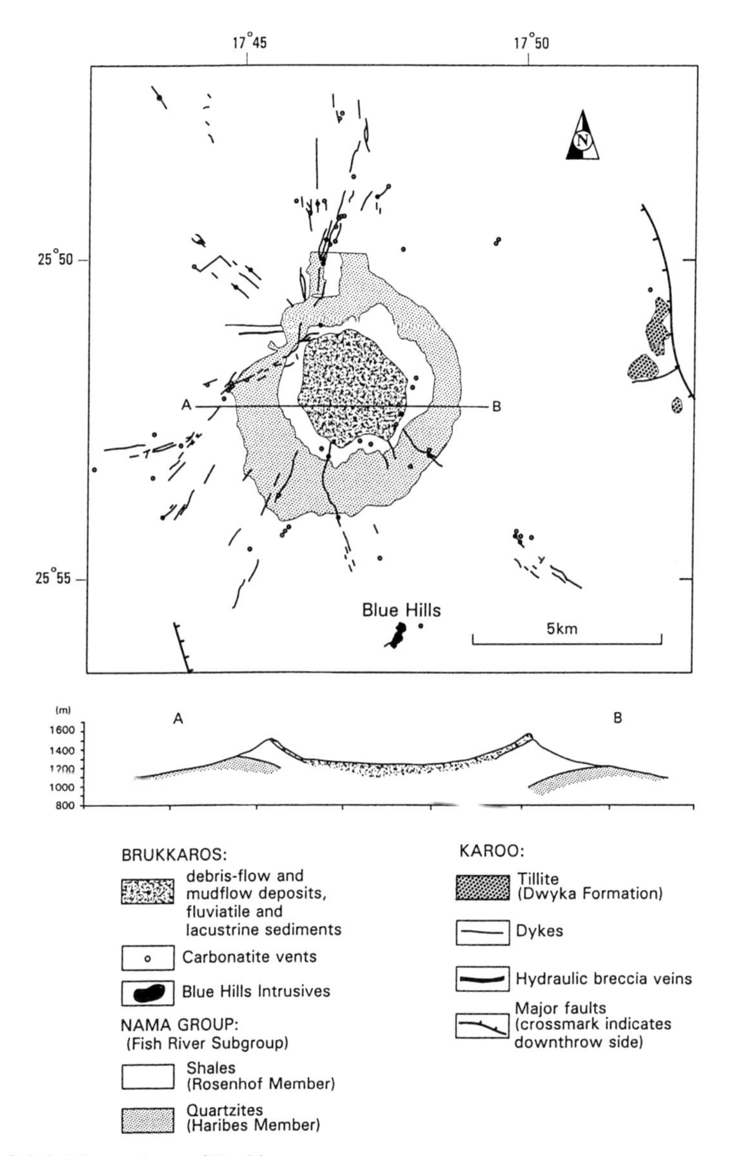


**Fig. 8.2.1:** The geology of Brukkaros.

younger, unconsolidated Kalahari sands and fluvial beds today form a regionally almost flat surface.

However, during Cretaceous time about 75 Ma ago, carbonatite-rich magma intruded the basement rocks and uplifted all the overlying Nama and Karoo beds (Kurzlaukis 1994). This uplift caused radial fissuring of the large up-domed area and allowed carbonatite magma to be emplaced within fissures and dykes. At those localities where the hot magma intersected groundwater, it exploded forming carbonatite pipes and maar craters up to possibly 1 km in diameter and surrounded by a low rim of ejecta. However, that the dykes reached the original surface and erupted by degassing, forming carbonatite scoria cones, and even lava flows, is conceivable, but cannot be proven, because all potential evidence has since been eroded.

This withdrawal of the magma from the reservoir at depth within the granites and gneisses and its rise inside the dykes and pipes towards the surface led to a mass deficiency in the magma reservoir. This mass deficiency caused the rock sequence of the roof of the reservoir to collapse into the reservoir with the consequence that a large caldera formed. Subsequently, groundwater and meteoritic water filled the caldera to produce a crater lake. Debris from the maar crater ejecta with carbonatite ash grains and lapilli and regionally eroded material was washed from all sides into the crater lake, thereby increasingly filling it up. After the volcanic activity came to an end, the magma remaining in the reservoir at depth cooled and released hot gases into the overlying rocks and groundwater. This hot groundwater and gases altered rocks of the central part of the Brukkaros caldera. The Brukkaros beds within the crater became silicified and quartz, amethyst and locally also smoky quartz crystallized in fissures and vugs. In addition, barite and calcite crystals formed (Stachel et al. 1994, 1995).

Uplift of southern Africa during Cretaceous and Tertiary times led to an extensive lowering of the landscape by regional erosion. In the area of the plain surrounding Brukkaros about 650 m of Kalahari, Karoo and Nama beds were eroded in several stages, unlike the extremely hard silicified Brukkaros beds which were originally laid down in a crater below the surface. At the highest elevation of Brukkaros a remnant of the regional peneplain is partially preserved but is itself already estimated to be 50 to 100 m below the regional surface which had existed at the time Brukkaros formed (Stachel et al. 1994).

## 8.3 Burnt Mountain

by Dougal Jerram and Gabi Schneider

At Burnt Mountain in southwestern Damaraland, a sheet of dolerite, associated with the Etendeka volcanism (see 8.6), intruded mudstones of the Verbrande Berg Formation, Karoo Sequence. The succession at Burnt Mountain (Fig. 8.3.1) starts of with the Verbrande Berg Formation, consisting of mudstones with coal horizons and sandstone channels. The Verbrande Berg Formation is overlain by coarse fluvial sandstone of the Tsarabis Formation, followed by calcareous mudstone of the Huab Formation. Lacustrine red mudstones of the Gai-As Formation rest on the Huab Formation, and the succession is capped by the aeolian Etjo Sandstone.

The mudstones were originally deposited in a lacustrine environment, and therefore contained carbonaceous material derived from remnants of organisms. During contact metamorphism caused by the intrusion of the hot dolerite magma, the organic material evaporated from the sediments, leaving behind a black, clinker-like rock, mainly composed of fritted clay minerals. A secondary coating of manganese minerals adds a purple luster to the black rocks. In contrast, the sediments show all shades of white, yellow and red, and this dramatic combination of colours, coupled with the lack of vegetation, has resulted in the name "Burnt Mountain".

## 8.4 Dieprivier

by John Ward

The prominent Conas Cliffs on farm Dieprivier (Fig. 8.4.1) W of road C14 S of Solitaire are composed of reddish-brown sandstone that forms part of the Tertiary age Tsondab Sandstone Formation – an extensive, composite sedimentary deposit that underlies much of the modern main Namib Sand Sea between the Orange and Kuiseb Rivers. Those sands, now consolidated to sandstone, accumulated under desert conditions some 10 to 20 Ma ago. On farm Dieprivier, the Conas Cliffs form part of the thickest known sequence of these ancient desert deposits. Water boreholes indicate thicknesses of up to 220 meters in places.

The internal structure and shape of the sandstone units exposed in the Conas Cliffs demonstrate that deposition took place largely in a desert

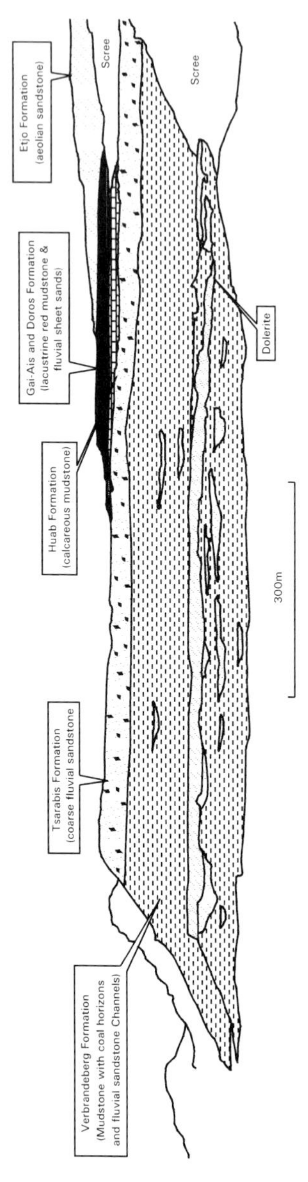


**Fig. 8.3.1:** Geological section through Burnt Mountain.

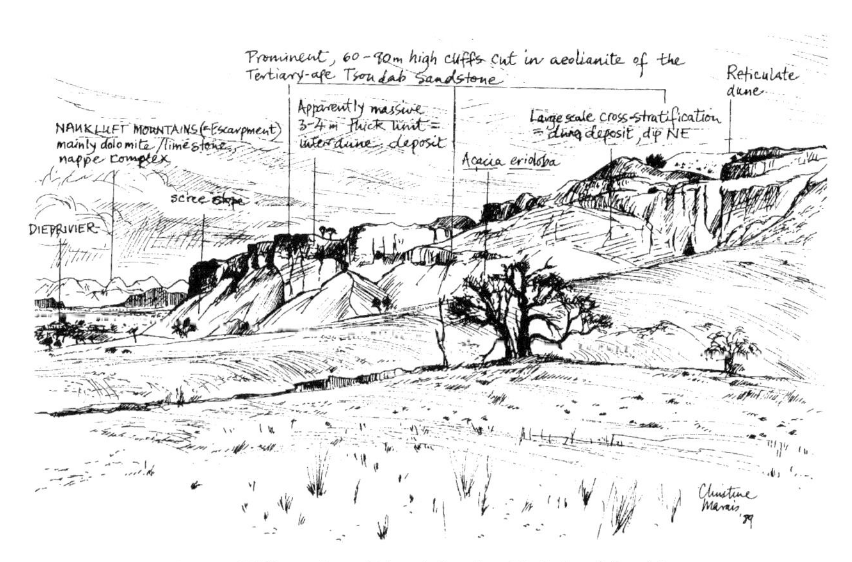


**Fig. 8.4.1:** The Conas Cliffs on farm Dieprivier (by Christine Marais).

dune and sand sheet environment. In the lower cliff section, large-scale cross-stratified units dip up to 30° to the NE - indicating a palaeo-wind regime with a dominant S component. Higher up the section, the influence of northwesterly and easterly winds is apparent from the orientation of the cross-bedding in those aeolian deposits. The large size and orientation of the aeolian dune bedding preserved in the Conas Cliffs suggests that large star dune forms were present along the Namib margin during those ancient desert times.

The much thinner, 2 to 5 m thick, flat-lying and structure-less beds represent interdune sand sheet accumulations, which were particularly favoured settings for Tertiary desert life. This biological activity is recorded by trace fossils – root-like features and the remnant trackways of termites, beetle larve, other invertebrates and an organism that left a trail remarkably similar to that of the modern golden mole which today is endemic to the Namib Desert dunes. The Conas Cliffs were carved by water erosion during the development of the Dieprivier drainage system which itself is a tributary of the ephemeral Tsondab River.

## 8.5 Dolerite Hills

by Volker Lorenz

In the surroundings of Keetmanshoop, widespread sub-volcanic rocks of the Keetmanshoop Dolerite Complex characterise the landscape. The age of the dolerite complex is Early Jurassic, about 180 Ma. It covers an area of more than 18000 $km^2$ extending for about 170 km in a N-S direction and for about 110 km from E to W. The Keetmanshoop Dolerite Complex consists of two major intrusive sills, and a number of dykes. It forms part of an even larger complex which extends partly under cover southward to the Orange River.

The sills are basaltic sheets; their magma mostly intruded parallel or in step and stair fashion along bedding planes of bedded sediments. Thick sills formed during the intrusion through inflation from originally thin sills, and simultaneous uplift of the overlying sediment cover. Intrusion into the sediments led to chilling of the hot magma by the cold sediments, and resulted in finely crystalline and partly glassy rocks along the sill base and top contact zones. The interior of the thick sills released its heat much more slowly, whereby larger crystals formed and consequently the interior of the thick sills are usually much coarser grained.

Dykes form when the Earth's crust is under tension and vertical or subvertical fissures open up through which magma rises towards the Earth's surface. In flat lying bedded sedimentary rocks dykes cross-cut the bedding. Similar to sills, thick dykes form via inflation and the thin marginal facies of dykes is finegrained to glassy whereas the voluminous interior facies is coarse grained.

In the Keetmanshoop Dolerite Complex, the sills and dykes consist of olivine tholeiitic rocks, which consist of plagioclase, clinopyroxene, olivine, and the accessory minerals apatite, magnetite and ilmenite. Their geochemistry is very similar to the geochemistry of the lava flows of the Kalkrand Basalt Formation further N. The dykes in the Keetmanshoop area in fact have been suggested to represent feeder dykes of the lava flows of the Kalkrand Basalt Formation. This, despite the fact that there is a distance of more than 200 km between the present day outcrops of the dykes and lava flows. In support of this assumption small and large subsided blocks of dolerite sills occur in the pipes of the Brukkaros Volcanic Field and in some of the pipes of the Gibeon Kimberlite Field. This relationship suggests that dolerite sills intruded that far N but regionally became eroded.

The sediments into which the sills and dykes intruded are part of the Dwyka and Ecca Groups of the Karoo Sequence. The lower sill has intruded into Dwyka Formation sediments and has a thickness of 150 m, the upper sill intruded into Ecca Formation shales and has a minimum thickness of 120 m. Erosion of the overlying sediments has widely exposed the two sills, especially the upper sill west, N and NE of Keetmanshoop, and to a minor extent also towards the S. The lower sill forms the hills just N of the Löwen River, where the upper sill has already been eroded completely. The lower sill is also exposed in an area extending from the Löwen River northeastwards towards Garinais. Due to the arid climate, the dolerite does not weather easily and remains rather hard. The softer Dwyka and Ecca Formation sediments of the Karoo could therefore be eroded more easily than the harder dolerite sills and the lower and upper sill usually cap hills and extensive plateaus in the larger Keetmanshoop area.

The exposed dolerite sills display spheroidal weathering which led to the formation of small and large dolerite boulders strewn over the hilly surface of the sills and their vicinity. One such site is at the famous Quiver Tree (Kokerboom) Forest on Farm Gariganus (Fig. 8.5.1), about 23 km NE of Keetmanshoop, where weathering of the dolerite of the upper sill has led to such typical boulders, so that this site is also known as "Giant's Playground". The Kokerboom Forest consists of approximately 250 quiver trees, *Aloe dichotoma*, that make use of the fertile soil formed from the dolerite sill, whereas in the surrounding sandy plains there are hardly any quiver trees. At the Kokerboom Forest, not only the trees can be marveled at, but also the dolerite boulders of the upper sill and locally the underlying baked, finegrained sediment of the Ecca shales (Reid 1998).

The dolerite dykes are typically 3 – 6 m wide, they rarely exceed 10 m. The length of the dykes varies from several 100 m up to more than 20 km for a dyke just east of the main road 50 km N of Keetmanshoop. Most dykes occur in a NW-SE trending zone mostly E of the main road N of Keetmanshoop. Spheroidal weathering of the dykes also led to the forma-

**Fig. 8.5.1:** Dolerite boulders at the Quiver Tree Forest, farm Gariganus.

tion of large and small boulders strewn over the dykes and their vicinity, which therefore frequently form distinct boulder ridges (Gerschütz 1996).

## 8.6 Etendeka Plateau

The Etendeka Plateau of volcanic rocks occurs in northwestern Namibia covering a region of some 78 000 $km^2$, with the main outcrop located between Cape Cross and the area south of Sesfontein, and outliers at Gobobosebberge, Albin, Huab, Khumib and Sarusas (Fig. 8.6.1). The volcanics of the Etendeka Group have an age of 132 Ma. They reach a maximum preserved thickness of 880 m at Tafelberg (Fig. 8.6.2) in the SE, while the original maximum stratigraphic thickness is believed to be in excess of 2

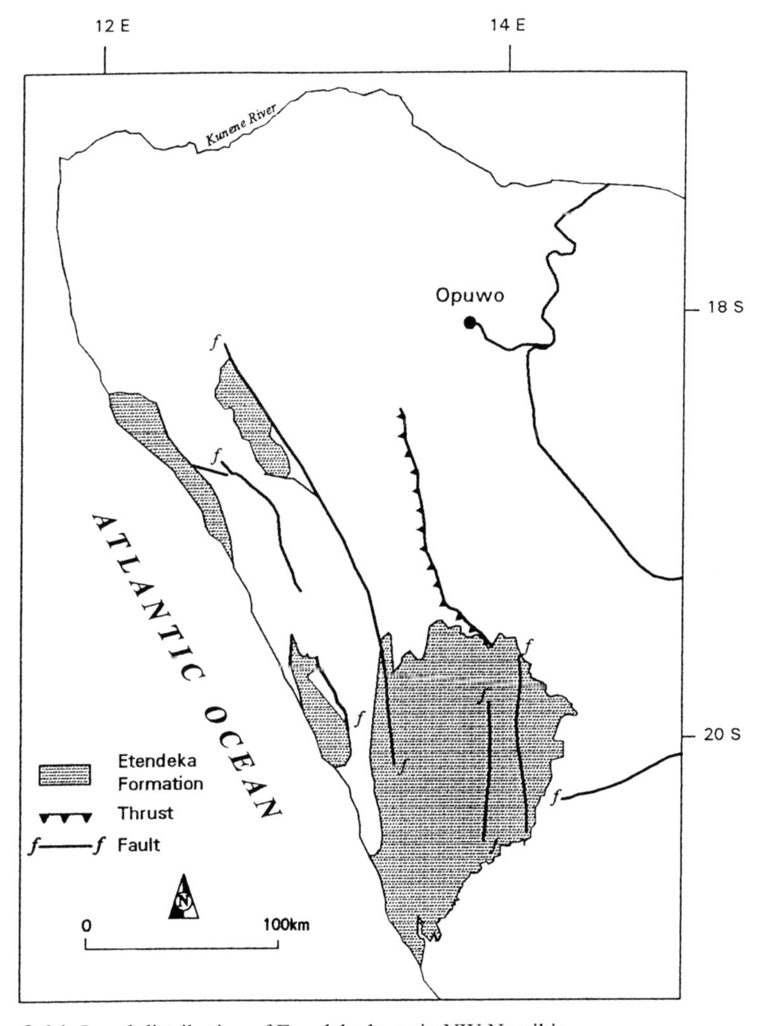


**Fig. 8.6.1:** Local distribution of Etendeka lavas in NW Namibia.

km (Reuning & Martin 1957). The landscape is characterised by high, table-topped mountains, which gave the area the Himba name "Etendeka",

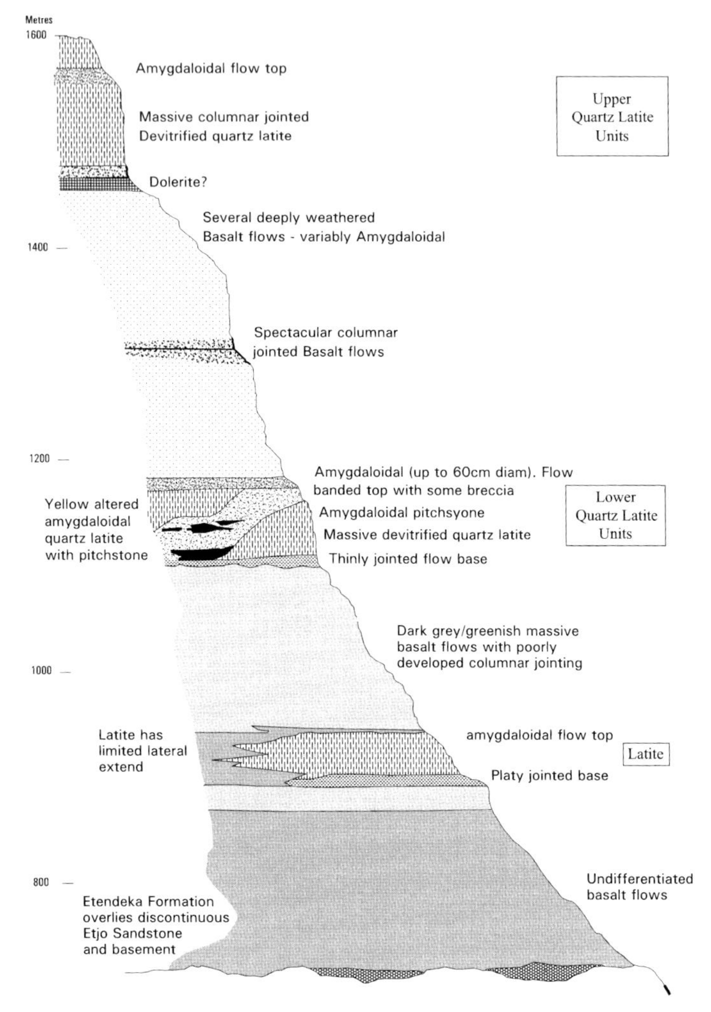


**Fig. 8.6.2:** Stratigraphic section through Tafelberg (after Milner et al. 1988).

meaning place of flat-topped mountains. Rocks of the Etendeka Plateau can be correlated with the volcanics of the Paraná Basin of Brazil, together with which they formed a major igneous province in western Gondwanaland just before continental break-up. This province had an estimated original volume of up to 2 million cubic kilometers of volcanic rocks, and is one of the largest continental flood volcanic provinces in the World.

Volcanics of the Etendeka Group consist of a bimodal association of mafic to intermediate (51–59 % $SiO_2$) tholeiitic lavas interbedded with more felsic (66–69 % $SiO_2$) quartz latite rheoignimbrites. The basalt flows and the massive quartz latite units at higher stratigraphic levels overlie the aeolian sandstones of the Etjo Formation of the Karoo Sequence, and frequently also overstep the Karoo Sequence to rest on older, Damaran basement (Fig. 8.6.2). The basal basalt flows are typically up to 40 m thick and are frequently intercalated with lenticular Etjo sandstone lenses, which then form conspicuous yellow bodies within the grey to reddish lavas. Quartz latites represent a significant portion of the succession, and constitute about 20 % of the rock outcrop. They form widespread units between 40 and 300m thick. Dolerite dykes and sills associated with the Etendeka Formation often intrude basement rocks, but rarely Karoo strata and the overlying volcanics.

Five basaltic lava types can be geochemically distinguished, namely the Tafelberg, Albin, Khumib, Huab and Tafelkop Basalts. The Tafelberg Basalts are by far the most abundant basalt type in the main Etendeka lava field. They are interbedded with the Khumib Basalts north of Terrace Bay and become progressively rarer further N. The Huab Basalts are a very fine-grained variety. The Tafelkop Basalts occur at the base of the succession in the northern part of this volcanic province and are the only olivine-bearing basalts in the sequence.

The quartz latite units appear to represent individual flows and have features common to both lavas and ash-flow tuffs. They can be divided into basal, main and upper zones. The main zone usually constitutes over 70 % of the unit and is fairly featureless. The basal and upper zones of the flow are up to 6 and 10 m thick respectively, and are characterised by flow banding, pitchstone lenses and breccias with pyroclastic textures. These characteristics enable the discrimination of individual flows. Several of the units are affected by zones of alteration resulting from hydrothermal activity and degassing within the flow shortly after deposition. These flow-top breccias are commonly mineralised by quartz, agate, zeolite and calcite.

There are three main quartz latite horizons named the Goboboseb, Springbok and Tafelberg Quartz Latites. The Goboboseb Quartz Latite is the oldest, consists of three flow units and crops out along the southern margin of the lava field. The Sprinbok Quartz Latite is best developed just N of the Huab River. Its upper part is truncated by erosion and is transgressively overlain by Tafelberg Basalts (Milner et al. 1988). The Springbok Quartz Latite is composed of upper and lower members which are separated by approximately 90 m of basaltic lavas. The base of the lower member is situated approximately 150 m above the base of the Etendeka Formation, and the contact with the underlying basalt is transgressive. The upper Springbok Member has a maximum thickness of 270 m and thins out rapidly towards the N. Outcrops of the more massive upper Springbok quartz latites are characterised by large, rounded, boulder-strewn hills, which are in contrast to the typical trap-basalt outcrops of the Tafelberg volcanics. Upper and lower quartz latites of the Tafelberg succession are up to 100 m thick and are separated by some 320 m of basaltic material. The volcanics of the Tafelberg succession exhibit a classic trap morphology and individual flows with a constant thickness can be traced for many kilometres. The Tafelberg Quartz Latite forms the thick uppermost layer of the flat-topped mountains of the eastern Etendeka Plateau south of Palmwag (Fig. 8.6.2) (Milner 1986).

## 8.7 Erongo

by Markus Wigand and Rolf Emmermann

The Erongo Complex with a diameter of approximately 35 km is one of the largest Cretaceous anorogenic complexes in northwestern Namibia. It represents the eroded core of a caldera structure with peripheral and central granitic intrusions (Emmermann 1979, Pirajno 1990). Surrounding the outer granitic intrusions of the Erongo Complex is a ring dyke of olivine dolerite, which locally reaches some 200 m in thickness and has a radius of 32 km. The ring dyke weathers easily and is therefore highly eroded. However, it can be easily identified on aeromagnetic data and satellite images. The best exposure is where the road from Omaruru to Uis cuts the ring dyke, about 12 km W of Omaruru. Here, one finds large, rounded blocks of the dark green or brown olivine dolerite. Compositionally similar NE striking basaltic dykes are also abundant in the vicinity of this locality (Fig. 8.7.1). Erongo is also the locality where well-known German geologist Hans Cloos undertook his fundamental studies on granite published in his book "Gespräch mit der Erde" (Cloos 1951).

The central part of the Erongo complex consists of a layered sequence of volcanic rocks, which form prominent cliffs rising several hundred meters above the surrounding basement. The interior of the complex is deeply eroded, giving access to the roots of the structure. The basement rocks consist of mica schists and meta-greywackes of the Kuiseb Formation and various intrusions of granites. In the SE, the rocks of the Erongo Complex overlie the Triassic Lions Head Formation, which consists of conglomerates, gritstone, arkose with interbedded siltstone and mudstone, and quartz arenite.

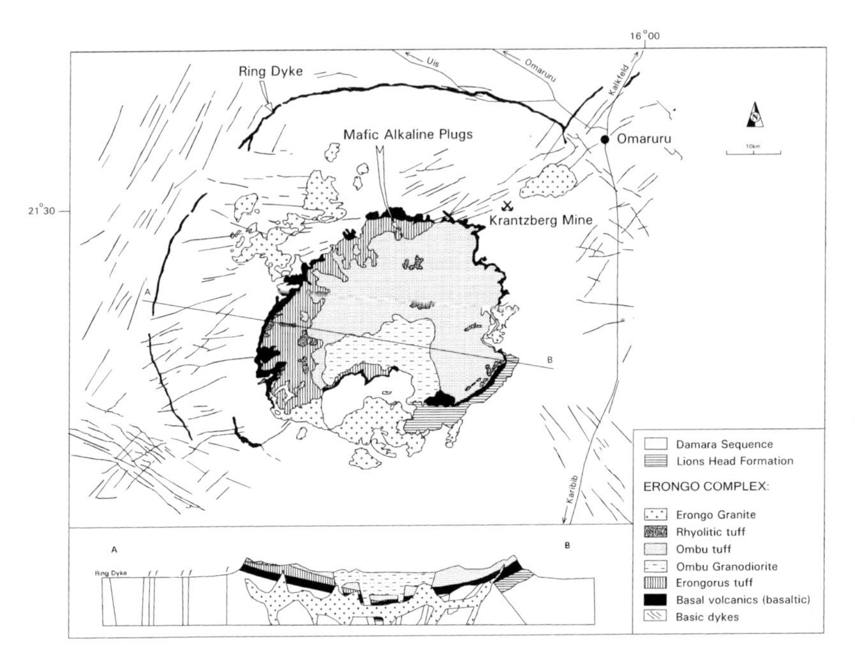


**Fig. 8.7.1:** The geology of Erongo.

The base of the Erongo Complex consists of a series of flat-lying basaltic lava flows and interbedded pyroclastic rocks. These basal volcanics are exposed throughout the entire complex and may originally have had an even wider distribution. With some 300 m thickness, the thickest layers of the basal volcanics are located in the southeastern part of the complex. The rock compositions range from tholeiitic, fine-grained basalt and basaltic andesite to andesite. Most basalts are considerably altered and commonly amygdaloidal, with vesicles filled with calcite and chalcedony. The alteration also caused local growth of quartz, actinolite, epidote and chalcopyrite. Plagioclase is sericitised and saussuritised, and clinopyroxene is replaced partially or fully by chlorite and epidote.

The basal volcanics are followed by a sequence of felsic volcanic units, which have been subdivided by Pirajno (1990) into four phases. The first phase is characterized by minor eruptions of mafic-intermediate lavas and major ash-flow tuffs of intermediate to felsic composition. These so-called Erongorus ash-flow tuffs occur mainly in the N, the NW and the W of the complex, but the units are absent in the E. They are generally altered and characterized by a devitrified groundmass with phenocrysts of quartz and altered K-feldspar.

The Erongorus tuffs are overlain by the Ombu ash-flow tuff sequence, which is volumetrically the main rock type of the complex and forms the most prominent cliffs. The Ombu tuffs are generally more quartz-rich than the Erongorus tuff units, but the most striking difference is that the Ombu tuffs frequently contain abundant sizable (cm- to dm) fragments of basement rocks. Compositionally, the Ombu tuff units are of rhyodacitic and rhyolitic composition.

The next volcanic phase was an explosive eruption and intrusion of rhyolitic magmas. Rhyolites occur as erosional remnants of rheomorphic tuff at the top of the volcanic sequence, and rhyolite dykes are especially common in the northeastern part of the complex.

The final volcanic to sub-volcanic phase involved the intrusion of subvolcanic plug-like bodies of basanites, tephrites and phonolites, which intrude Erongorus tuffs at the northern edge of the complex. All rocks are under-saturated in silica having less than 5 % normative nepheline. A good, and possibly the thickest exposure of the basal basalts occurs on Farm Niewoudt on the south-eastern edge of the complex. On the way to the farmhouse one passes exposures of Ombu tuff and can also see the contact between the basalts and the conglomerates of the Lions Head Formation, Karoo Sequence, and the contact between the Erongo basalt and the Ombu ash-flow tuff.

Intrusive into the basal volcanics is the Ombu granodiorite. This granodiorite body is mineralogically and geochemically very similar to the Ombu ash-flow tuff rocks and contains the same distinctive assemblage of basement xenolith fragments. The contact between the two units is gradational. Based on these observations, Emmermann (1979) concluded that the Ombu ash-flow tuff units are the erupted equivalents of the granodiorite. The Erongo granite is volumetrically the most important intrusive phase of the Erongo Complex, and is found as isolated stocks, dykes and sills around the complex. The granite is a massive, coarse-grained, equigranular leucocratic biotite granite with an average modal composition of 36 % quartz, 33 % perthitic orthoclase, 25 % albite, 4,5 % biotite and 1,5 % accessory minerals (Emmermann 1979). Finer-grained facies, aplitic dykes and rare pegmatitic pods and lenses also occur. Accessory minerals include tourmaline, beryl, zircon, monazite, fluorite, apatite and topaz. An exceptional and distinctive feature of the Erongo granite is the presence of quartz-tourmaline "nests", up to 30 cm in diameter, which occur in all facies of the granite. The "nests" consist of tourmaline, quartz, K-feldspar, plagioclase, biotite, fluorite, apatite, topaz and cassiterite. On the farms Ameib and Omandumba one can find all forms of Erongo granite. An interesting formation of Erongo granite is the group of huge boulders known as "Bull's Party" at Ameib. Rock paintings are common in the Erongo granites, and the Phillips Cave on farm Ameib is a National Monument.

Like many high-level granites, the Erongo granite is associated with mineralisation. Tourmalinisation is widespread and greisen-type tungsten, tin, fluorite and beryllium mineralisation is found locally. By far the most important mineral deposit is the Krantzberg tungsten deposit near the northeastern margin of the complex. The Krantzberg Mine was a major tungsten producer in Namibia and it is estimated that approximately 1 million tons of ore were extracted before closure in 1980 for economic reasons. Ferberite, fluorite, cassiterite, beryl, as well as molybdenum-, iron- and copper-sulphides are common ore minerals at Krantzberg. This mineralisation appears in replacement-type greisen rocks and quartz-tourmaline breccias. The mineralisation took place when boron- and fluorine-rich hydrothermal fluids induced extensive selective replacement of pre-existing

Damaran granites by quartz, sericite, topaz and tourmaline. The formation of the hydrothermal fluids has been connected with the emplacement of the Erongo granite (Pirajno & Schlögl 1987, Pirajno 1990).

Geochemical variation diagrams show that the felsic volcanic and intrusive units of the Erongo Complex can be derived from a common magma source. Emmermann (1979) suggested that the Ombu tuffs and granodiorite were derived from crustal rocks similar to those in the local basement because of their peraluminous composition, the fact that they contain cordierite, and the abundance of crustal xenoliths. New evidence in support of crustal origin comes from a radiogenic isotope study of the Ombu granodiorite and the Erongo granite (Trumbull et al. 2001). Both units cover a narrow range of values, which overlaps completely with data from Damaran metasediments and granites. However, the basaltic rocks at the base of the Erongo are geochemically distinct from the felsic units, they must therefore have a different source, and by analogy with the chemically similar Etendeka flood basalts elsewhere in Namibia, their source is thought to be the upper mantle.

## 8.8 Etosha Pan

by Gabi Schneider and Pete Siegfried

The Etosha Pan is situated in the southern part of the Owambo Basin. This basin is an intracontinental sedimentary basin floored by Mesoproterozoic rocks of the Congo Craton. It contains some 8000 m of sedimentary rocks, the uppermost of which belong to the Kalahari Group sediments of the northwestern outlier of the huge Kalahari Basin of southern Africa (Miller 1997) (see 2.13). Etosha Pan itself covers an area of 4760 $km^2$, and has a maximum N-S extent of 80 km and an E-W extent of 120 km. The pan has almost no relief, with altitudes between 1077 to 1085 meters above sea level.

The uppermost unit of the Kalahari Group, the Andoni Formation, forms the bedrock of the pan floor. The Andoni Formation consists of silt and sand. In places, the Andoni Formation is overlain by the Etosha limestone. Transitions between the two formations, with the Etosha limestone becoming more sandy, do occur. Only a very shallow upper part of the profile of Etosha Pan is influenced by recent surface waters altering the rocks of the Kalahari Group. This part has a mineral assemblage typical of a saline-alkaline environment, including analcime, K-feldspar, sepiolite, saponite, calcite, dolomite, strontianite and various salts (Buch & Rose 1996). It is noticeable that large parts of the surface of the pan are coloured green-gray by the micaceous mineral glauconite.

The geological history of the Etosha Pan is characterised by the formation of a palaeo-lake followed by erosion (Fig. 8.8.1). The palaeo-lake was formed through a drainage system including the upper Kunene and Kavango Rivers in the Late Miocene, some 5 to 7 Ma ago. This lake was fully established some 3 Ma ago and covered an area of 55 000 km. If it existed today, it would be the third largest lake in the World. However, a westward flowing river eroded its way inland and captured the headwaters of the

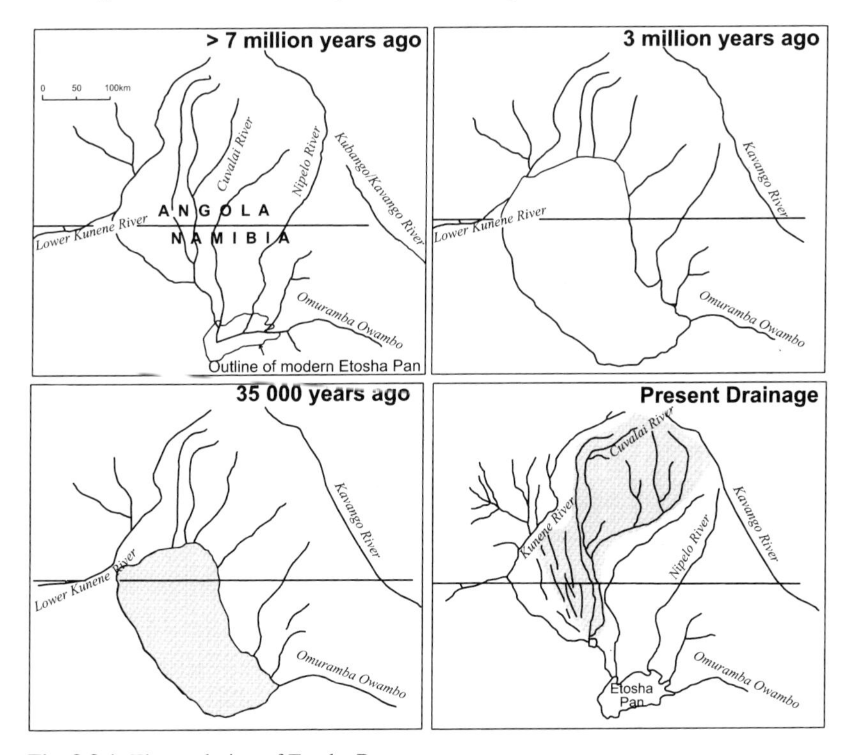


**Fig. 8.8.1:** The evolution of Etosha Pan.

Kunene River so that by the end of the Pliocene it no longer supplied "Lake Etosha" with water. The size of the lake diminished through evaporation, and the Cuvelai System with oshanas (small seasonal rivers) developed in its place. Pleistocene stromatolites, which are sedimentary structures formed by carbonate precipitation as a result of the growth and metabolic activity of microorganisms, are also present. This indicates a quiet lacustrine environment supersaturated with carbonates, as can be expected from an evaporating inland body of water (Smith & Mason 1991). Once the lake had vanished due to the dry climatic conditions, a pan developed which is only supplied seasonally with water through rainfall and flood events, locally named efundjas, in the oshanas (Marsh & Seely 1992).

After the Kunene River's headwaters were captured, "Lake Etosha" was also no longer supplied with fluvial sediments. A change in the palaeoclimate to more arid conditions at the end of the Pliocene resulted in erosional processes during the Quaternary being characterised by fluvio-lacustrine processes in the rainy season with aeolian deflation processes during the dry season. This development of Etosha Pan as an erosional structure during the Quaternary was effected by the seasonal swamping of parts of the pan's floor and the associated disintegration of the underlying sedimentary rocks of the Kalahari Group. During the dry season, the winds transport the disintegrated sediments out of the pan and form the prominent dunes to the NW of the pan. These erosional processes lead to the continuous denudation of the pan floor since the Late Pliocene/Early Pleistocene (Buch 1996).

A number of springs occur along the southern margin of Etosha Pan, and provide water for game. The source of the water are the dolomite mountains of the Damara Sequence to the S of the pan. Karst structures in these mountains provide ample mobility for ground water, and where the dolomites are in contact with the younger, clay-rich sediments of the Kalahari Group, springs develop, as the clay is impermeable.

## 8.9 Fishriver Canyon

The Fishriver Canyon of southern Namibia is the second largest canyon in the World after the Grand Canyon of the USA. It consists of a northern upper and a southern lower canyon. From the first waterfall north of the northernmost viewpoint, to a point opposite the Chudaub trigonometrical beacon, the canyon is 56 km long, measured along the river course. The lower canyon is between 460 and 550 m deep and 5 km wide, whereas the upper canyon is only 160 to 190 m deep, but 8 km wide.

A proto-Fishriver Canyon 300 m above the present day level of the river and floored by sandstone and black limestone of the Kuibis Subgroup of the Nama Group formed as early as 300 Ma ago, since at the end of the Dwyka glaciation sediments of the Dwyka Group were deposited here. Erosion was directed mainly laterally, and the peneplain of the Huns Plateau formed.

Major continental uplift, following the break-up of Gondwanaland some 130 Ma ago, resulted in the deep incision of the river to its present day level. Judging from the numerous meanders, the river must have flowed very slowly on a very even surface before its dramatic incision. At first, it cut through the horizontal layers of the Kuibis Subgroup rocks, but later reached the underlying gneisses, amphibolites and migmatites of the

**Fig. 8.9.1:** The Fish River Canyon.

Namaqualand Complex, which was metamorphosed some 1200 Ma ago. These metamorphic rocks are cut by dark coloured, linear diabase dykes, which are about 770 Ma old. There is a very obvious contact between the Nama sediments and the massive Namaqualand Complex rocks, which show a very different erosional pattern and display a typical metamorphic texture (Fig. 8.9.1).

The upper canyon is a tectonic trough bordered to the W by a sharp monoclinal fold, and on the eastern side by a linear reverse fault. The lower canyon also exposes a number of steeply inclined bounding faults, some of which are well visible from the main lookout point. These faults also affect Karoo Sequence rocks, and must therefore be younger than 300 Ma. While the canyon without doubt formed in a rift valley, some of the faults actually indicate compressional tectonics (Schmidt-Thome 1981). Groundwater is using these faults to reach the surface in the form of hot springs, of which Ai-Ais with 60 °C is the most famous. The so-called sulphur spring upstream is also well-known and has a temperature of 56 °C.

## 8.10 Gamsberg

The flat-topped Gamsberg Mountain is a well-known landmark in southern central-western Namibia just S of the Gamsberg Pass. It is part of the Great Escarpment that separates the highlands in the east from the low-lying Namib Desert in the west. The enormous massif, with a total elevation of 2347 m above sea level, towers some 450 m above the Khomas Hochland in the N and E, whereas to the W, the difference in altitude to the Namib plain is 1100 m. Due to its form and height, it is the most prominent mountain in the area and visible from a distance of more than 100 km.

The largest part of the mountain consists of a reddish, coarse-grained to porphyritic granite. This granite intruded rocks of the Gaub Valley Formation, Rehoboth Sequence, as well as the Weener quartzdiorite, some 1100 Ma ago. It was then deeply buried by Damaran sediments. Damaran deformation at about 540 Ma ago gave the granite its typical strong, NW dipping foliation. Erosion over the next 150 Ma removed the Damaran rocks, but left the hard granite of the Gamsberg as a relatively higher region with a flat erosive top. The absence of lower and middle Karoo strata on top of the granite suggests that the Gamsberg formed an island during the onset of Karoo sedimentation some 300 Ma ago. However, towards the end of the Karoo period, some 180 Ma ago, when the whole of southern Africa was covered by a vast desert, the difference in elevation was reduced to such an extent that aeolian sandstones of the Etjo Formation were deposited on the Gamsberg granite. Today these silicified sand dunes form a 30 m thick layer of hard sandstone, and this capping has protected the underlying granite from subsequent erosion during various stages of uplift (Fig. 8.10.1) (Schalk 1982).

The Etjo sandstone is capping numerous inselbergs in northern central Namibia, but this occurrence is by far the southernmost, and is situated more than 200 km S of the others. It covers an area of approximately 3 km$^2$, and consists of a lower, 1 m-thick brown sandstone layer, overlain by a thick layer of whitish, fine-grained sandstone with frequent cross bedding. An interesting feature of Gamsberg is the presence of numerous sandstone-filled fissures, which may have resulted from earthquakes during Karoo times. The seismic activity must be seen in connection with the

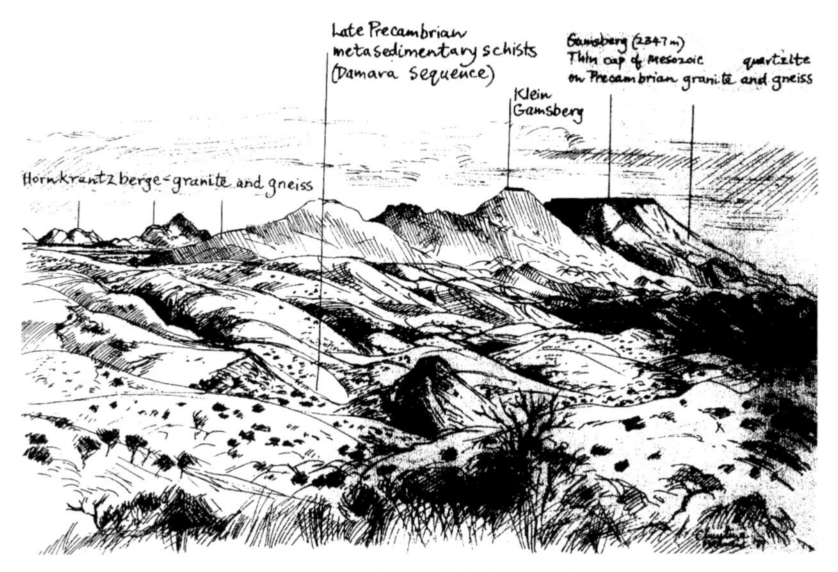


**Fig. 8.10.1:** The Gamsberg from the NE (by Christine Marais).

beginning break-up of Gondwanaland, during which time the area of the Gamsberg was affected by strong extensional trends in an E-W direction leading to the formation of fissures striking in a predominantly N-S direction. During this time, the Etjo sandstone was not yet solidified, and aeolian sand could therefore infiltrate the fissures, which are up to 10 cm wide and penetrate the granite to a depth of more than 200 m (Wittig 1976).

## 8.11 The Gibeon Meteorites

The Gibeon Meteorite Shower is the most extensive meteorite shower known on Earth and covers a large elliptical area of some 275 by 100 kilometers centered on Brukkaros S of Mariental. Most fragments are located just southeast of Gibeon and to date, some 120 specimens with a weight of almost 25 tons have been recorded. Unfortunately, an unknown number have been collected unrecorded.

Small pieces of metal were collected by James Alexander in 1838 at a place "about three days journey NE of the mission at Bethany". He described the pieces to be up to two feet square, and sent some material to the chemist John Herschel in London, who established the meteoritic origin of the material. However, it is well known that the local Nama people had been using the meteorites for a long time to produce spear points and other weapons.

The first recorded large piece, weighing some 81 kg was carried by oxwagon for 800 miles to Cape Town by John Gibbs before 1853 (Shepard 1853) and from there it was sent to London, where the mineralogist to Queen Victoria, Professor John Tennant, purchased it. He forwarded it via New York to Professor Charles Shepard of Amherst College in Massachusetts, who studied the material in detail. At least ten pieces of Gibeon meteorites had been shipped to Europe by 1910 (Marvin 1999), before Dr. Paul Range, geologist of the German Colonial Administration, collected all the remaining meteorites he could find and mapped their occurrence between 1911 and 1913. The specimens were displayed in Windhoek, and a number of them were donated to various museums around the world.

The Gibeon meteorites occur partially embedded in rocks of the Karoo Sequence and calcrete of the Kalahari Group. It has been calculated, from available evidence, that the Gibeon meteorites resulted from a meteorite body, measuring roughly 4 by 4 by 1.5 meters entering the Earth's atmosphere along a northwesterly trajectory and at a low angle of 10° to 20° from the horizon. This body fragmented while still high in the atmosphere, so that the fragments themselves suffered thermal alteration by melting of the outer surface. This melting either resulted in smooth outer layers, or in molten material being pulled off in places by the drag of the atmosphere, leaving behind an uneven mass with deep, spherical cavities on the outer surface. These well developed thermal alteration structures prove that the fragments had an extended flight through the atmosphere before being deposited. The varieties of thermal- and shock-induced microstructures of the Gibeon meteorites is the greatest in the world and only matched by the Canyon Diablo Meteorite of northern Arizona (Marvin 1999).

The Gibeon meteorites are classified as octahedrites, the most common type of iron meteorite, and consist entirely of taenite and kamacite, two different crystalline varieties of an iron-nickel-alloy. Taenite (gamma-Fe with 8 – 55 % Ni) and kamacite (alpha-Fe with 5,5 % Ni) both have a cubic crystal lattice and form alternating bands orientated parallel to octahedral planes. This structure is known as Widmannstätten pattern, after its discoverer, Baron Alois von Widmannstätten, and is a characteristic feature of many meteorites (El Goresy 1976). Besides iron, which is the main constituent, the Gibeon meteorites contain an average of 8 % nickel, 0,5 % cobalt, 0,04 % phosphorous, small amounts of carbon, sulphur, chromium and copper, and traces of zinc, gallium, germanium and iridium. The Gibeon meteorite specimens range in size from one ton to just a few grams. A few rare crystals forming tetrahedrons and octahedrons have also been recovered, and probably resulted from the successive splitting off of thin kamacite lamellae along with the selective corrosion of intervening taenite layers (Marvin 1999).

Today, the larger known Gibeon meteorites that have remained in Namibia, are displayed in the Post Street Mall (Fig. 8.11.1) and at the museum of the Geological Survey of Namibia in Windhoek, and a number of smaller ones form part of the reference collection of the Geological Survey of Namibia. While meteorites continue to be found in the area of Gibeon, it is very difficult to locate them in the field without a metal detector. Meteorites are protected by law in Namibia, and may not be removed from their original site. Despite this, a large number continues to leave the country illegally to be sold at international rock and mineral shows. Recently, the

**Fig. 8.11.1:** The meteorite display in Windhoek's Poststreet Mall.

largest known specimen so far, weighing about one ton, was illegally removed and exported to the USA (Norton 1998). This deprives the local people of an opportunity to develop a sustainable use out of this unique natural resource and national heritage, and also deprives the international scientific community of valuable research material.

## 8.12 The Hoba Meteorite

The Hoba Meteorite is situated 20 km W of Grootfontein and represents the largest single meteorite known in the world today (Fig. 8.12.1). It was found by Jacobus Hermanus Brits in 1920 and was declared a National Monument on 15 March 1955. When first found, only a small portion of the meteorite was visible on surface. Early ideas of recovering nickel from the Hoba Meteorite were soon abandoned, since it did not prove to be eco-

**Fig. 8.12.1:** The Hoba Meteorite.

nomic. Some samples were removed using an oxyacetylene torch, leaving a long scar on one side, and others were taken from some drill holes. In 1954, a curator of the American Museum of natural History in New York attempted to purchase the meteorite to display it together with other large specimens in the USA. However, the railway line, which passes within 7 km of the Hoba Meteorite only, had an axle loading capacity of 24 tons only, hardly enough to transport a 60 ton meteorite! Another major obstacle was the transport from its location to the railway line. Through this good fortune (albeit unlucky for the curator), the meteorite remained in Namibia and is now one of its most prominent natural national monuments. Over the years, people continued to try to cut off samples from the meteorite, and in 1985, Rössing Uranium Ltd made funds available to the National Monuments Council to establish an information center and to protect the meteorite from further vandalism.

The Hoba Meteorite is situated on the northwestern edge of the Kalahari plain, underlain by calcrete of the Kalahari Group. Underlying the calcrete are Palaeoproterozoic granites of the Grootfontein Complex, as well as dolomites and limestones of the Otavi Group, Damara Sequence. Surprisingly, no crater or altered rocks have been found associated with the impact site and the survival of the meteorite as a single mass suggests that it entered the atmosphere on a long trajectory at a sufficiently low velocity to allow a soft landing. After the meteorite fell, it was gradually covered by layers of calcrete. This calcrete was formed by evaporation of near-surface groundwater, which contained calcium carbonate derived from the surrounding Otavi Group limestone. Today, the region receives a maximum annual rainfall of only 750 mm, and near-surface groundwaters are less abundant. The presence of calcrete therefore suggests a more humid climate in the recent geological past.

The meteorite weighs approximately 60 tons and measures 2,95 by 2,84 meters with a thickness between 1,22 and 0,75 meters. The shallow pits and depressions on the horizontal upper surface of the meteorite are the result of melting of the outer surface during the passage through the Earth's atmosphere. Minor oxidation has subsequently occurred on the surface, and is responsible for the green stains of secondary nickel minerals present on the surface of the meteorite. The meteorite is separated from the surrounding calcrete by a 30 cm thick layer of magnetic, dark-brown iron shale composed largely of limonite, with minor magnetite and trevorite [$NiFe_2O_4$]. This layer contains iron, nickel and cobalt in the same proportions as the meteorite itself, indicating that the oxidation proceeded in situ without any leaching away of components. Before oxidation, the meteorite could have weighed about 88 tons (Marvin 1999).

The Hoba Meteorite consists of 82,4 % iron, 16,4 % nickel and 0,76 % cobalt. Other elements present in traces are carbon, sulphur, chromium, copper, zinc, gallium, germanium and iridium. The Hoba Meteorite is therefore classified as an ataxite. Under the microscope, material from the Hoba Meteorite displays the typical compact ataxitic structure with faint lines, wedges and patches. The main minerals are kamacite (a nickel-iron alloy with 5 – 7 % nickel) and taenite (a nickel-iron alloy with up to 65 % nickel). Intergrowths of kamacite and taenite needles form the typical Widmannstätten structure (Fig. 8.12.2). The Hoba Meteorite also contains the rarer meteoritic minerals schreibersite [(FeNi)P3], troilite[[FeS] and daubreelite [$FeCr_2S_4$]. McCorkell et al. (1968) measured a significantly high content of a rare, radioactive nickel isotope, and assumed that this represented less than one half life of the isotope, suggesting that the Hoba Meteorite fell to Earth less than 80 000 years ago.

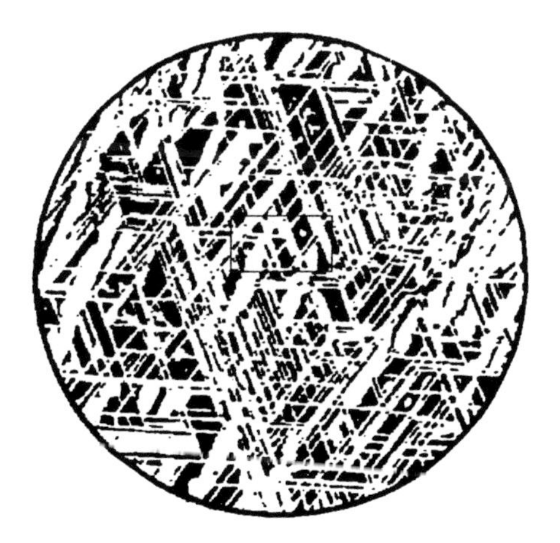

**Fig. 8.12.2:** The pattern of Widmannstätten structures.

## 8.13 The Kalahari

by Pete Siegfried

The Kalahari is one of the World's most extensive bodies of sand extending from the northern Cape as far N as the Congo River and from Namibia eastwards to Zimbabwe (Fig. 8.13.1). This huge area is characterized by often red, but sometimes white, sand overlying up to 300 m of variably ce-

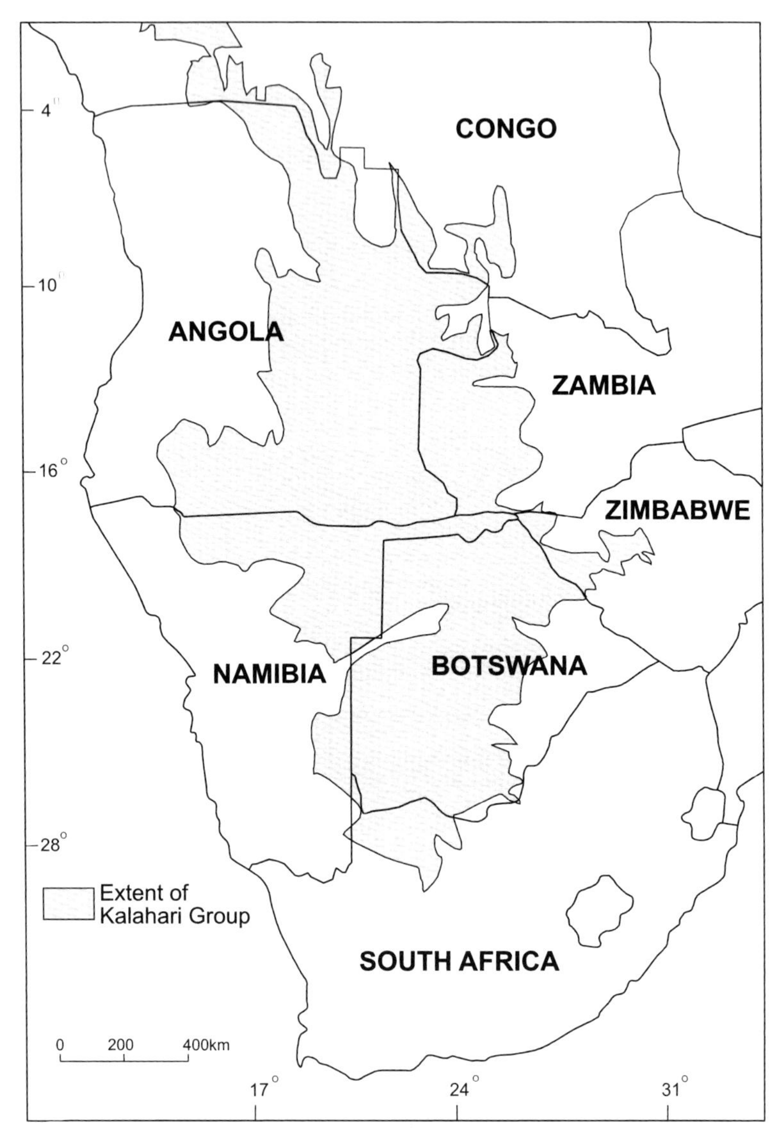


**Fig. 8.13.1:** Regional extent of the Kalahari Group.

mented sands, clays and gravels. The Kalahari is regarded as a thirstland not because it has an unusually low rainfall, but because any precipitation rapidly sinks into the sand, and for most of the year surface water is extremely rare. As the sediments are flat-lying and essentially sandy, not many exposures of rocks occur. Most of the knowledge about the sediments of the Kalahari has been derived from boreholes drilled for water and to a limited extent from observations around pans and along river cuttings.

The Kalahari is the most recent large-scale geological landform deposited in southern Africa and overlies Karoo sediments and volcanics. Its geomorphology is characterised by dunes, fossil drainages and pans. The drainages, which flow into central southern Africa, are referred to as omuramba in Namibia.

The Kalahari sand is generally 80 to 100 m thick, although there are many areas where sand in excess of 300 m has been found. It is often cemented into calcrete, a white, hard limestone material commonly used to construct the gravel roads in Namibia and calc-silcrete. Calcrete characteristically forms where evaporation potential is far in excess of precipitation. It is characteristic of all arid areas.

The age of the Kalahari is difficult to determine for various reasons. Fossils are very rare and the process of calcretisation tends to destroy pollen grains, which could be used to indirectly date these sediments. However, dates from fossil water, pan sediments and calcrete carbon isotope analysis indicate Quaternary ages, but deposition of Kalahari sediments probably started in the Late Cretaceous, soon after Gondwana broke up some 130 Ma ago. When South America split away, the edges of the African continent started to rise, causing preferential drainage to the interior of the subcontinent, where erosion did not provide for a passage to the sea. The Kwando and Kavango Rivers with their attendant inland deltas are present-day examples of this. There is evidence along the Orange River and elsewhere that the region was much wetter at the time, with extensive rivers and frequent floods transporting large sediment loads to the inland basin. The dune forms that are so characteristic of the Kalahari are believed to date back to at least the last glacial period, 16 000 to 20 000 years ago, and have since been stabilized and are therefore always vegetated.

## 8.14 Karas Mountains

The Little and Great Karas Mountains are situated S and SE of Keetmanshoop, where they rise about 500 and 1000 m respectively above the surrounding plains which are about 800 m above sea level. The highest peak in the Great Karas Mountains, the Schroffenstein, attains an elevation of 2202 m above sea level. The topography of the Karas Mountains is the result of extensive block faulting which occurred some 500 Ma ago, when the rocks making up these mountains were uplifted. Rejuvenation of movement along the faults and further upliftment of the already established mountain belt occurred some 135 Ma ago (Münch 1974). Since then, only erosion has modified the form of the Karas Mountains.

The basal unit of the Karas Mountains is the Namaqualand Complex, comprising sedimentary and intrusive rocks metamorphosed about 1200 Ma ago. These rocks are dissected by the so-called Gannakouriep Dyke Swarm, a number of altered (saussuritised and uralitised) hornblende-diorite dykes with an age of about 770 Ma. The overlying Nama Group is represented by quartzites of the Kuibis Subgroup, shales and quartzites of the Schwarzrand Subgroup and shales and quartzites of the Fish River Subgroup. The Dwyka Group at the base of the Karoo Sequence has been developed in places and is represented by a tillite considered to be a consolidated ground moraine (Fig. 8.14.1) (Genis & Schalk 1984).

## 8.15 Kolmanskop

Kolmanskop, the first railway siding to the E of Lüderitz and today a ghost town, is the place where it all began: The first authentic Namibian diamond was discovered here in 1908 by railway worker Zacharias Lewala while shoveling sand against the embankment of the railway line. He picked up a small stone with an unusual sparkle and handed it to his foreman August Stauch. Stauch took the stone to the hospital at Aus, where the medical superintendent Dr. Peyer tested it with hydrofluoric acid and identified it as a diamond. Stauch, having found that the thin coarse gritty sands between bedrock outcrops in the area also contained similar stones, pegged off large areas, but hardly anybody took any notice (Levinson 1983).

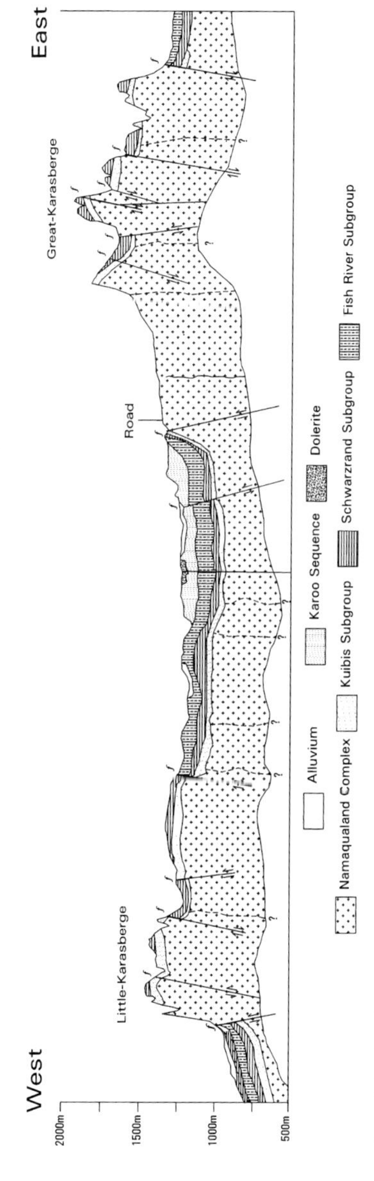


**Fig. 8.14.1:** Geological section through the Karas Mountains.

However, two months later, when Government geologist Dr. Paul Range confirmed the stones to be diamonds, the rush began, and in a short period of time practically the whole coastal strip from Lüderitz to 28° S was pegged. Within 18 months the whole coast from Walvis Bay to the Orange River mouth had been prospected (Wagner 1914). Some deflation surfaces S of Lüderitz, such as the Idatal and Scheibetal, were so enriched by wind action that the workers simply picked up the diamonds while crawling along the surface on their stomachs. By 1913, 20 % of the World's output of diamonds came from the area extending from Hottentot's Bay to 25 km S of Bogenfels (Boise 1915).

After the first rush had subsided, five major companies emerged, namely the "Kolmanskop Schürfgesellschaft mbH", the "Vereinigte Diamant-Minen AG", the "Diamanten AG", the "Deutsche Diamanten Gesellschaft" and August Stauch's "Koloniale Bergbau Gesellschaft". Diamond fields in the area were given German names such as "Schmidtfeld", "Emiliental", "Friedlicher und Feindlicher Nachbar", "Stettin", "Lübeck", "Bismarckfelder", "Zillertal", "Weilburg", "Großer Kurfürst" und "Rangefeld". At "Schmidtfeld", the first sand bucket dredger was employed and a small plant with a capacity of 18 cubic meters of sediment a day was erected in 1911.

Early treatment of the diamondiferous gravels started with the use of crude handwashing machines, and later swinging sieves were gradually replaced by trommel sieves. To facilitate better recovery, a large plant belonging to the "Koloniale Bergbau Gesellschaft", the remains of which can still be seen today, was installed S of Kolmanskop in 1910. Water was supplied to the plant from a large pumping station at Elizabeth Bay through a 25 km-long pipeline. Power was generated by the large power station erected by the "Koloniale Bergbau Gesellschaft" in Lüderitz, which can be seen on the seaward side of the road leading to the Nest Hotel. A narrow railway line, 20 km long, was built to transport the ore to the plant. Initially, the cocopans were pulled by mules, but electric locomotives were later introduced and the line was extended to 70 km. In total, more than 3 million carats of diamonds were recovered in the Lüderitz-Kolmanskop area prior to 1920. Mining continued at Kolmanskop until early 1931, when the mine closed due to the worldwide recession (Schneider & Miller 1992).

The name Kolmanskop, Kolmanskoppe or Kolmanskuppe originated from a transport driver by the name of Johnny Kolman. When transporting goods from Keetmanshoop to Lüderitz, he usually gave his oxen a rest and made camp in the vicinity of the gneiss hillock overlooking Kolmanskop. In 1905, he was caught in a heavy sandstorm, and all his oxen and his wagon vanished. He himself was fortunate enough to be rescued, and the hillock was subsequently named after him (Levinson 1983).

During extended geological periods the same heavy sandstorms were responsible for the relative concentration of diamonds near the surface, with all the lighter minerals been blown away. Aeolian placer or deflation deposits occur in a major deflation belt up to 15 km wide and extending from Lüderitz in the N to Chamais Bay in the S. Within this belt, salt-assisted weathering and aeolian abrasion have eroded elongated depressions up to 120 m deep into the bedrock consisting of Neoproterozoic Gariep Complex and older granitic gneisses of the Namaqualand Metamorphic Complex. These depressions parallel the dominant southerly and southeasterly winds which control the aeolian processes in the deflation belt. The winds frequently average 50 to 60 km/hour and gust at 80 km/hour. Sand is mainly supplied to the deflation belt from south-facing re-entrant bays along the coast. This sand forms trains of barchan dunes migrating northwards up to 60 m per year. Granules of all sizes and composition are rapidly sorted by size, density and shape, with heavy minerals deposited along wind barriers in the depressions, resulting from big boulders or crosscutting dykes (Corbett 1989).

As diamond mining progressed near Kolmanskop, a unique little settlement developed in the desert. Wooden buildings with corrugated iron cladding, pre-fabricated in Germany, gave way to impressive stone buildings which to this very day contrast in a bizarre way with the desolate desert surroundings (Fig. 8.15.1). Especially impressive is the double-storey house of the mine manager. Kolmanskop had every amenity one could think of, including a police station, a post office, a general dealer, a bakery, a butchery, a lemonade and soda water factory, an ice factory, a carpentry shop, stables, a primary school, a swimming pool, a hall with a stage and a bowling alley (Levinson 1983). Kolmanskop became the headquarters of "Consolidated Diamond Mines" (CDM), when the company was formed by Sir Ernest Oppenheimer in 1920 incorporating a large number of the diamond mining companies previously operative in the area. In 1943, however, the headquarters of CDM moved to Oranjemund, since mining operations in that area had become more important, and in 1956, the last resident left Kolmanskop.

**Fig. 8.15.1:** Old German house at Kolmanskop.

Kolmanskop has since become a ghost town, where one can see how the desert reclaims land previously taken by man. Nevertheless, a visit to Kolmanskop still tells the story of the diamond rush. One can marvel at the efforts made by humans in the middle of the desert and imagine the fortune, generated by diamonds, spent to create a German village in this remote and inhospitable area of Namibia.

## 8.16 Kuiseb Canyon

by John Ward

In the geomorphological scheme of the central Namib Desert, the Kuiseb Canyon is a relatively young feature, dating back to the Pliocene some 5 Ma ago. The Kuiseb drainage, the most prominent watercourse cutting westwards across the Namib Desert between the Orange River in the S and Walvis Bay in the central region, forms the spectacular northern boundary for much of the main Namib Sand Sea. The deep canyon incised by this ephemeral drainage system, in places reaching over 200 m deep, was initiated by relative base level changes due to continental uplift and sea level fluctuations over the last 5 Ma. At Kuiseb Pass, the river has cut its bed deeply into the schists of the Neoproterozoic Kuiseb Formation of the Damara Sequence (Fig. 8.16.1).

The incised course of the Kuiseb River also preserves a partial record of the style of that drainage system prior to the erosion downwards into the canyon. At the lookout point some 10 km W of Kuiseb Pass, traces of the pre-incision, Tertiary river are demarcated by boulders and smaller gravels that constitute the Karpfenkliff Conglomerate Formation. These fluvial and alluvial deposits have been cemented by calcium carbonate to form a resistant cap on older, water-laid arenites that are part of the Tsondab Sandstone Formation – itself a precursor to the current Namib Desert (see 8.21) deposited some 10 to 20 Ma ago.

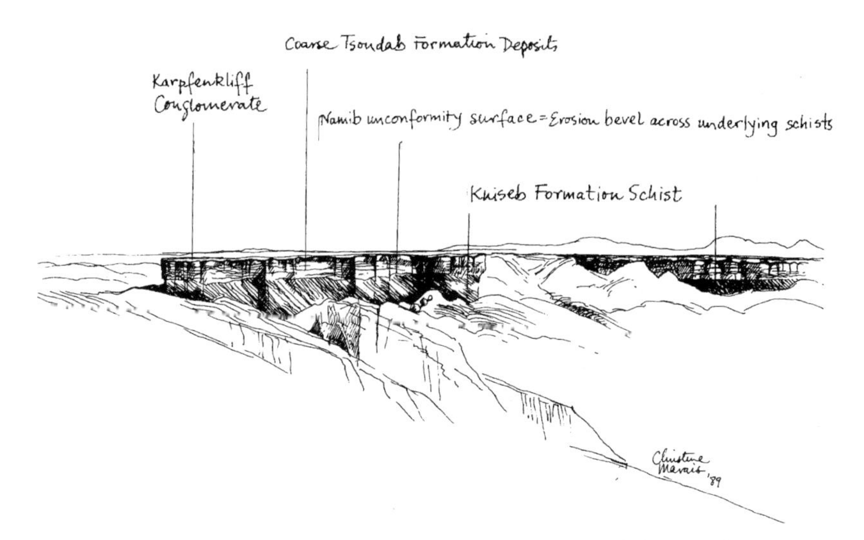


**Fig. 8.16.1:** The Gomkaeb Cliffs in the Kuiseb Canyon (by Christine Marais).

## 8.17 Lake Otjikoto and Lake Guinas

by Dieter Plöthner

Lake Otjikoto lies some 20 km W-NW of Tsumeb along the main road to Ondangwa and Namutoni. "Otjikoto" is the Herero word for "deep water", and is in fact a collapsed sinkhole in dolomites of the Neoproterozoic Maieberg Formation of the Damara Sequence (Hedberg 1979). It's location possibly has some structural control, as a magnetic anomaly passes adjacent to the lake. Lake Guinas is also located in bedded dolomites with associated structural control. It is larger than Lake Otjikoto, having a maximum diameter of 140 m and a depth of 153 m (Penney et al. 1988). The waters of Lake Otjikoto and Lake Guinas are utilised for irrigation (Hoad 1992, 1993).

Sinkholes commonly occur in areas underlain by carbonate rocks, and form through dissolution of the carbonates by meteoric waters. Such dissolution often starts from underground, and the roof of the structure eventually collapses, leaving behind a circular depression. The Karstveld to the northwest of Tsumeb shows a number of sinkholes and may be traced from the Tsumeb-Grootfontein-Otavi Triangle westward through the Etosha National Park and from there northwards. Cavities in dolomitic rocks form when water percolates through fractures and carbonic acid in the water dissolves the rock. Gradually these cavities grow into caves that usually contain water. In some cases the roof of the caves collapses to form open sinkholes, such as Otjikoto and Guinas. The lakes at Otijkoto and Guinas represent "windows" to the surface of the groundwater flow from the Otavi Mountain Land towards the N.

The depth of the Otjikoto sinkhole possibly exceeds 75 m. It has a diameter at surface of about 100 m surrounded by vertical cliffs, but becomes wider at depth (Fig. 8.17.1). The storage of the lake is of the order of 400 000 m$^3$, and the Government permitted maximum abstraction for irrigation purposes amounts to 1.3 million m$^3$ per year (Hoad 1992; 1993). In the period 1973 to 1980 the lake water level actually rose by 12 m reflecting groundwater recharge due to a number of above mean annual rainfall years. Since 1980, however, the water level has been almost continuously declining, as most of the annual rainfall has been below average and groundwater recharge extremely poor. This drought scenario was observed in the entire Otavi Mountain Land and its surroundings (Schmidt & Plöthner 1999).

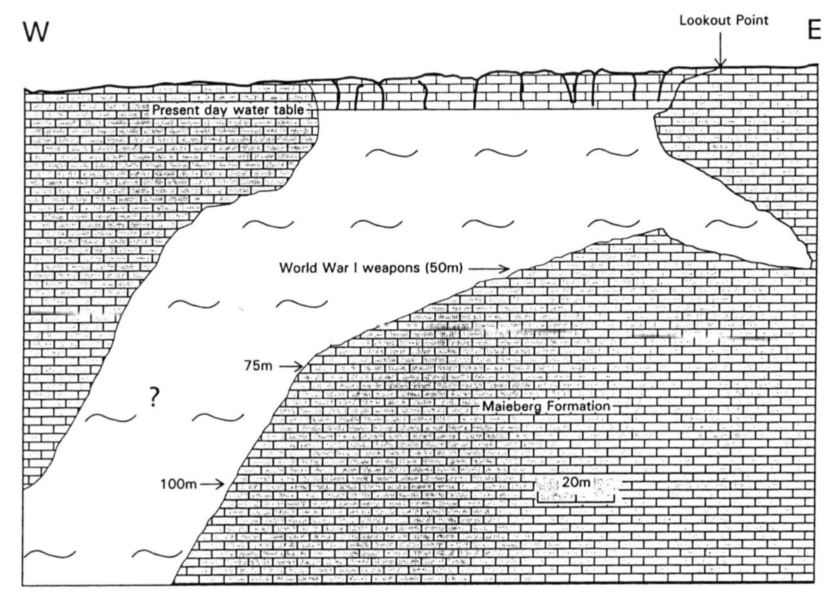


**Fig. 8.17.1:** Geological section through Lake Otjikoto.

The water of Lake Otjikoto can be characterised as a magnesium-calcium-bicarbonate water, which has a low mineralisation of 470 mg/l, a pH value which varies between 7,0 and 8,7 in different parts of the lake and a temperature of 27 °C (Marchant 1980). The predominance of magnesium over calcium is typical for dolomite groundwaters.

The water level was first measured by Sir Francis Galton in 1851, and Galton was also the first to record the presence of fish, which were later to be found new to science, in Lake Otjikoto (Galton 1889). In 1933, the English scientist and explorer Dr. Karl Jordan collected fish specimens from the Otjikoto and Guinas lakes and made them available for formal scientific description (Trewavas 1936), including two native species, *Tilapia guinasana* and the southern mouthbrooder *P. philander* (Skelton 1990).

Many changes have taken place since Galton and Jordan visited the two lakes. A number of man-made structures have been built including platforms for large pumps and water gauging installations. It is also well known that the retreating German forces during World War I dumped large quantities of weapons and ammunition into Lake Otjikoto in 1915. A large portion of that military equipment was recovered by the South African troops and later used in East Africa (Penrith 1978).

Early in the development of mining activities at Tsumeb, water required for processing was transported by ox-wagons from Lake Otjikoto to Tsumeb. In 1907, a high pressure pipeline was laid and the steam engine built by M. Neuhaus & Co., Luckenwalde to run the pumps for conveying 500 $m^3$/day of water from Lake Otjikoto to Tsumeb can still be seen on the southern side of the lake. In 1909, excessive rainfall resulted in a rise of lake water of 9 m in Otjikoto and some 30 m in Guinas in 1909 (Jaeger & Waibel 1920). As a consequence, the Otjikoto pump scheme was inundated and Tsumeb had to be supplied with water carried by railway from Otavi (Gebhard 1991).

Results of stable isotope determinations suggest that the water of Lake Otjikoto and Lake Guinas is being recharged from an area in the Otavi Mountain Land at an average altitude ranging from about 1600 to 1900 m above sea level. Such altitudes occur in the area S of Tsumeb and N of Kombat. Radiocarbon dating of the dissolved inorganic carbon of the lake water by Vogel (1978) resulted in $^{14}C$ water age ranges from 600 to 1600 years before present for Otjikoto and Guinas, respectively. Thus, the travel rate of groundwater from the mountainous recharge areas to the lakes, amounts to 30 to 40 m/year. This is a rate typical for groundwater circulating in fractured, but not highly karstified aquifers such as the prevailing dolomites of the Otavi Mountain Land and its surroundings (Ploethner et al. 1997)

A 1,3 m core of sediments from a sloping platform at about 52 m depth in the central part of Otjikoto Lake was obtained in 1990. The core consists of a grey homogeneous fine silt with small white and dark specks and vague white laminations between 44 and 65 cm. Attempts were made to date the core with radiocarbon. Conventional $^{14}C$ ages of 1300 and 3200 years BP at a depth of 20 and 94 cm, respectively, yield a sedimentation rate of approximately 0,5 mm/year (Scott et al. 1991). On the basis of pollen data the Otjikoto sequence indicates that during Late Holocene times, some 3500 years ago, a relatively dry climate prevailed, which gradually changed into a temporarily wetter phase probably about 1000 years ago, which is in good agreement with the palaeoclimatic concept of Heine (1992).

## 8.18 Messum

The Messum Complex of northwestern Namibia is one of a series of Cretaceous anorogenic igneous complexes and lies just to the SW of Brandberg. Messum distinguishes itself from the other Mesozoic complexes by its highly diverse assemblage of intrusive and extrusive rocks, which in turn appear to have intruded volcanic rocks of the Goboboseb Mountains. It has a diameter of 18 km and is a multi-stage gabbro-granite-rhyolite-syenite subvolcanic ring structure. The outline of the complex in satellite photographs is roughly circular, however, on the ground, the outer contact with the Goboboseb lavas is often obscured by Cenozoic sediments. Dating of the various rock units of the Messum Complex has established an age for their emplacement of between 132 and 135 Ma (Korn & Martin 1954, Milner et al. 1995).

The rocks of the Messum Complex can be divided into an outer series of gabbro and granite, and an inner core of syenites and rhyolitic breccia some 6 km in diameter. Two distinct ridges of basalt occur within the outer eastern and southern parts of the complex and represent the earliest subsidence phase of the intrusion (Fig. 8.18.1). The basalts are extensively intruded by dykes and veins of granite, which contain basaltic xenoliths.

Volcanic breccias and rhyolite lava domes form the central hills of the Messum Complex. The breccias and lavas have undergone contact-metamorphism and are intruded by syenite, dolerite and granite. They are also interbedded with large quartzite bodies, interpreted as re-crystallised Etjo sandstone of the Karoo Sequence, which underlies the Goboboseb volcanics. The volcanic lithologies of the central Messum Complex are interpreted as the earliest uplifted sequence in the history of the complex.

The oldest intrusions in Messum are best exposed in the eastern and southern parts of the complex and consist mainly of granite with abundant mafic inclusions and syenite. These are followed by two gabbroic intrusions, namely the eastern gabbro, which is the older of the two, and the western gabbro. The last major intrusion in Messum is formed by syenite intruding the inner core region of the complex.

The relationship between the different intrusive phases is complicated and provides evidence of extensive magma mixing, which in turn led to the development of the large number of different subordinate magmatic hybrid rocks, such as theralites, basanites and phonolites (Ewart et al. 1998). In

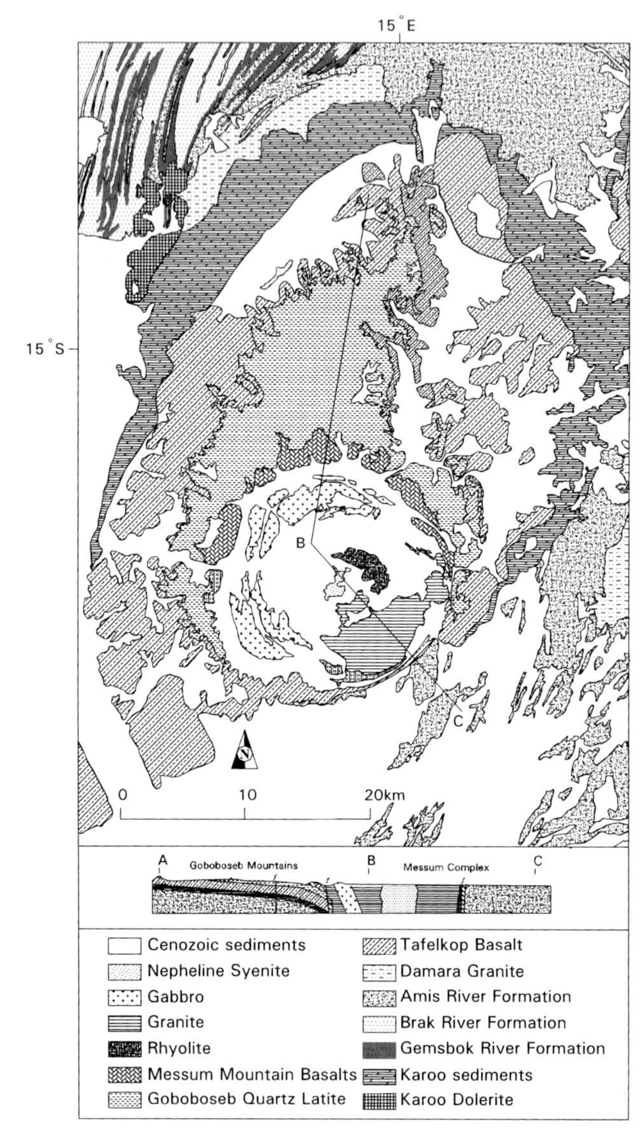


**Fig. 8.18.1:** The geology of Messum.

summary, the basalts are the oldest phase and represent the subsidence stage of the intrusion. The rhyolites are fragments of a subaerial volcanic superstructure built during an early phase in the evolution of a Messum volcano. The rhyolites appear to have collapsed along a caldera margin and have been partially engulfed in the rising syenite magma. The Messum gabbros predate the alkaline rocks of the core. They were intruded by syenites, and this event was accompanied by collapse of a large fragment of the superstructure of the Messum volcano. Then followed the last stage of the intrusion represented by the syenites of the core (Harris et al. 1999).

The relationship of the Messum Complex with the surrounding Goboboseb volcanics is of particular interest (Fig. 8.18.1). The Goboboseb Mountains are a southern remnant of the Etendeka Formation volcanics consisting of a 600 m thick sequence of quartz latites and basalts, which almost completely enclose the Messum Complex, and gently dip towards it. Field evidence and the close relationship between some of the Messum intrusives and the Goboboseb quartz latites, suggests that the Messum volcano may have been the source for some of the quartz latites. Messum might have played a prominent role as magma source during the formation of the Etendeka continental flood basalt, which is one of the largest continental flood basalt bodies in the world (see also 8.6). Like the Etendeka volcanics, the Messum Complex rocks have been dated at 132 Ma (Milner & Ewart 1989).

## 8.19 Mount Etjo

Mount Etjo is one of the more striking morphological features in northern central Namibia. It forms a long, narrow plateau, some 16 km in length with a maximum width of 3 km SE of Kalkfeld. The highest elevation of 2086 m above sea level is attained at the south-western edge of the plateau.

Mount Etjo is one of a number of Karoo age inselbergs composed of rocks that formed in the Omingonde Basin of northern central Namibia between 220 and 180 Ma ago (compare 8.23 and 8.32). Under semi-arid conditions, this basin accumulated up to 700 m of conglomerates, sandstones and mudstones, before the climate became extremely arid and the aeolian sandstones of the Etjo Formation were deposited. All these inselbergs occur just south of the Waterberg Fault, a prominent fault along which base-

**Fig. 8.19.1:** Mount Etjo.

ment rocks were thrust upward against Karoo rocks. The basement rocks partly protected the Karoo rocks from weathering, and today Mount Etjo is an erosional relic with a capping of 70 m of hard Etjo Sandstone (Fig. 8.19.1).

At Mount Etjo, the Lower Omingonde Formation, represented by sediments of a stratified inland lake, overlies granites and gneisses of the Damara Sequence. The Middle Omingonde Formation consists of gravel beds deposited by braided rivers, whilst the Upper Omingonde Formation accumulated under increasingly arid conditions, starting with sediments of loessic plains with saline lakes and ponds. These are overlain by gravel beds that were deposited by rivers meandering on a semi-arid floodplain, which in turn are overlain by the sand-beds of rivers meandering on semi-arid loessic floodplains with shallow saline lakes. The contact to the overlying Etjo Formation is marked by wadi sediments and barchanoid dunes, and further up in the sequence large longitudinal sand dunes with periodically wet interdune flats are encountered (Smith & Swart 2000).

**Fig. 8.19.2:** Skull of *Erythrosuchus africanus*.

Vertebrate fossils have been collected in abundance from Mount Etjo, including herbivores of various sizes and the large carnivore *Erythrosuchus* (Fig. 8.19.2) (see 5.). The dinosaur footprints at Otjihaenamaparero (see 8.25) occur in lower Etjo Formation sediments close to Mount Etjo.

## 8.20 Mukorob

Mukorob, the Nama term for "Finger of God", used to be a prominent landmark in southern Namibia (Fig. 8.20.1). The pinnacle consisted of an upper, top-heavy sandstone pillar some 12 m in height and 4,5 $m^2$ in area, resting on a thin neck of soft mudstone, some 3 m long and only 1,5 m wide. The sandstone head weighed about 450 tons. The neck was supported by a broader mudstone base, and the whole structure towered about

**Fig. 8.20.1:** Mukorob.

34 m above the surrounding plain. The collapse of Mukorob was reported on 8 December, 1988. Today, only the base and a part of the neck remain.

The sandstones and mudstones of the Mukorob are part of the Upper Carboniferous Mukorob Formation of the Karoo Sequence (Symons et al. 2000), which forms the prominent cliffs of the Weissrand Escarpment due E of Mukorob. The Mukorob Formation was deposited about 270 Ma ago. Following the continental break-up of Gondwanaland, rapid erosion occurred. This erosion, as a result of the incision of the Fish River, eventually produced the Weissrand Escarpment a few million years ago. Mukorob itself was an erosional relic, formed because the harder sandstone cap was a resistant layer protecting the mudstones and slowing their rate of weathering. It has been estimated from the fairly small distance of its position from the Escarpment that it became an isolated feature only 50 000 years ago (Miller et al. 1990).

The collapse of Mukorob occurred in the course of natural weathering. Through time erosional processes slowly reduced the size of Mukorob with the soft mudstone neck eaten away at a greater rate than the sandstone head. Finally, the neck became so thin that it could no longer support the weight of the head. Slaking of the mudstones of the neck during the rainy season further aggravated the process. A fracture plane inclined at 45° to the NE formed, and Mukorob, as it collapsed, slid down this inclined plane. The inclination of the plane caused the rock column to tilt, so that the mudstones of the neck fell on the northeastern side of the base, while most of the sandstone fragments from the top can now be found on the southwestern side.

An interesting aspect of the collapse of Mukorob is its possible association with a huge earthquake, which took place in Armenia on 7 December 1988. The seismological station of the Geological Survey of Namibia in Windhoek, with a distance from the epicenter of some 7500 km, recorded the earthquake waves during the morning of that day. While the major factor in the collapse of Mukorob remains the erosion of its neck, the earthquake waves may well have provided the final impetus for the mudstones to give in to the weight of the sandstone head.

## 8.21 Namib Desert

by John Ward

The Namib is one of five west coast, low-latitude desert regions in the world. It forms a narrow desert tract some 2000 km long and less than 150 km wide, lying W of the Great Escarpment and stretching northwards along the Atlantic seaboard from the Olifants River in South Africa, through the entire length of Namibia to about the Carunjamba River in southwestern Angola. The climate of the central section, lying wholly within Namibia, is extremely arid. The area N of the Kunene River is classified as an arid summer rainfall desert and that S of the Orange River as an arid winter rainfall desert.

The current desert conditions in the Namib are attributed to the interacting, aridifying effects of four phenomena. Namely the subtropical South Atlantic anticyclone, which is usually situated offshore between Lüderitz and the Orange river mouth; the cool northward-flowing Benguela Current and, more importantly, its associated cold-water upwelling system; the absence of convection with temperature inversion in the lower atmosphere related to the cool maritime influence, and the divergence of the South East Trades. These effects operate on the comparatively stable, passive margin setting of southwestern Africa, which has no deep, post-Gondwana break-up sedimentary basins onshore.

Geomorphologically, the Namib hosts an astonishing variety of landscapes, ranging from classic desert dunes and extensive gravel plains, in places with gypcrete and calcrete duricrusts, to dissected terranes called

**Fig. 8.21.1:** The Namib Sand Sea meets the Atlantic Ocean south of Sandwhich Harbour.

"gramadullas", ephemeral watercourses forming linear oases in a harsh environment, inselbergs, low mountain ranges and even the distal reaches of three perennial rivers, namely the Olifants, Orange and Kunene Rivers, that rise well outside the desert. Along the coast, wind-swept sandy beaches, rocky stretches – in places cliffed and carved into arches, sand spits and coastal pans are common components of the dynamic, Atlantic seaboard, which is characterised by strong southerly quadrant winds and foggy conditions (Fig. 8.21.1).

The main Namib Sand Sea – a 34,000 km$^2$ tract of sandy desert lying between about Lüderitz in the S and the Kuiseb River in the central region – is one of the more significant active desert sand and dune accumulations outside the Sahara. Although small by comparison to the Sahara, the main Namib Sand Sea hosts most of the dune types known in the world (Fig. 8.21.2). These dune types are distributed across this desert tract from west to east, reflecting the change from a strong, uni-directional southerly wind

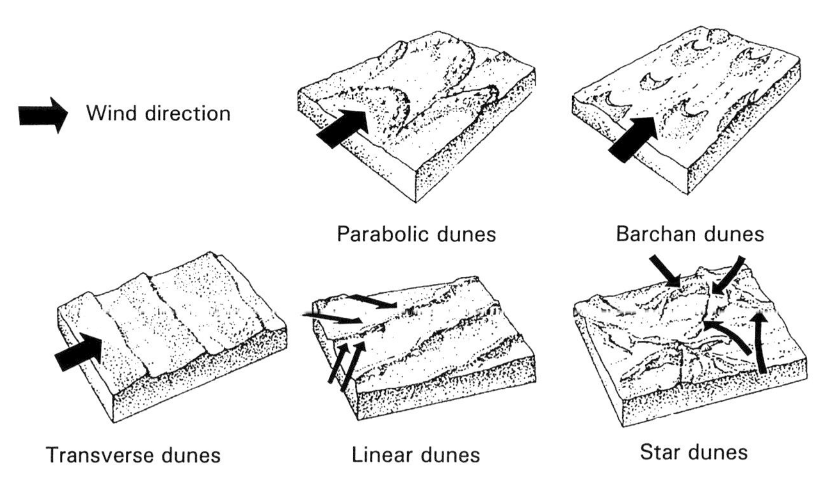


**Fig. 8.21.2:** Dune forms of the Namib Desert.

regime along the Atlantic coast to bi- and even multi-directional systems influenced by north-westerly and easterly winds further inland.

The immediate coastal tract is dominated by white to beige coloured, crescentic dune forms, which include 30 m high barchans travelling northwards at rates of about 50 metres per year and composite transverse dunes up to 75 m high. Within the central zone, orange coloured, complex and compound linear dunes, in places over 100 km long and 120 m high, form the core of the main Namib Sand Sea. The eastern margin is characterised by large stellate dunes (up to 220 m high) and lower, partly vegetated, network dunes that display a striking orange-red colour.

The age of the Namib Desert has generated considerable scientific debate. Recent fossil evidence, particularly from the southern and central Namib, is shedding more light on the rates and duration of accumulation of the extensive Tertiary desert deposits that underlie much of the main Namib Sand Sea. Following the break-up of Gondwana in the Early Cretaceous, erosional processes controlled the formation of the Great Escarpment and the development of the narrow coastal tract, and, in places, exhumed the inherited Gondwana relief. In the southern Namib, the Orange River was a large, mature meandering drainage that tapped much of the western hinterland of southern Africa. Uplift at about the Cretaceous – Tertiary boundary 65 Ma ago initiated the formation of the Great Escarpment in southern Namibia and promoted the incision of the Orange River. This latter process continued well into the Tertiary and possibly even to the present day, imparting a steep, westward-directed gradient to that drainage system. Silcrete duricrusts preserved in the southern Namib testify to geomorphic stability and a change from a climatic regime in which deep weathering and kaolinization was prevalent to drier conditions at about the Cretaceous-Tertiary boundary.

Initial indications for a proto-Namib desert phase occur in the Eocene, but this phase reached it's full extent in the Middle Miocene some 15 Ma ago. Coastal palaeo-dunes deposited under a southerly palaeo-wind regime associated with an Eocene shoreline in the southern Namib and evaporitic deposits of similar age in the northern coastal Namib indicate the onset of arid conditions in the Namib tract. Apart from minor alluvial deposits of Early Miocene age in the southern Namib, the Miocene was characterised by widespread terrestrial sedimentation under desert conditions in which the southerly quadrant wind regime was dominant (see 8.4). During this phase, extensive desert aeolianite deposits were laid down – the bulk of their sand having been derived from the Orange River sediments debouched into a wave-dominated delta with a strong longshore drift that moved material northwards under the influence of the southerly winds – a situation similar to that prevailing today. Along the lower Orange Valley, aggradational terraces hosting a rich fossil fauna were emplaced within the ancient meander loops of the inherited drainage.

The widespread desert conditions in the Middle Tertiary were terminated by a return to wetter climatic conditions that promoted extensive alluvial sedimentation off the Great Escarpment and off low mountain ranges and inselbergs in the desert during the Middle to Late Miocene. These high-lying calcareous gravels and conglomerates pre-date the deep canyon incision of many of the westward-directed drainages.

Following the alluvial activity, climatic conditions dried out during the Late Miocene and possible Pliocene, and geomorphic stability prevailed such that widespread calcareous soils were formed – giving rise to mature calcrete duricrusts that commonly cap both the Proto-Namib and older alluvial sediments in the Namib landscape. Although younger calcrete palaeosols are found in the Namib, the older profiles were developed prior to the deep canyon incision of the westward-directed watercourses that cut across the Namib tract.

Since the full establishment of the cold-water Benguela upwelling system in the Late Miocene some 5 Ma ago (Ward et al. 1983), terrestrial sedimentation in the Namib has been dominated by arid to hyperarid climates with a dominant southerly wind regime that was punctuated by short-lived wetter intervals. Consequently, the more spectacular desert deposits are the sand seas and dune fields of the Namib with pan carbonates and spring tufas recording the wetter interludes. During this phase, the larger westward-directed rivers rising east of the Namib cut deep courses across the desert with their terrace sequences recording several aggradational and degradational phases during the Quaternary. The regional hierarchy of these Late Cenozoic fluvial terraces – from older, coarse gravel deposits to finer-grained younger sediments – supports the contention that aridification in the Namib has been essentially progressive to the present day.

## 8.22 Naukluft

The Naukluft Mountains of central western Namibia represent an impressive mountain range, which forms part of the Great Escarpment. They occupy an area of 2100 km$^2$ and have their summit on farm Arbeit Adelt at an elevation of 1960 m above sea level. The flat, plateau-like top of the mountain complex is separated from the adjacent highland plateau to the S by the impressive near vertical cliffs of Johann Albrechts Felsen, while in the NW and in the W its highest peaks tower almost 1000 m above the plains of the Namib Desert. Together with vast parts of the desert, the Naukluft Mountains form the Namib-Naukluft Park, the name Naukluft being derived from the Afrikaans and German languages meaning "narrow valleys".

The Naukluft Mountains have a very interesting geological history. They are composed of three main geological units, the basement rocks of the Mesoproterozoic Rehoboth and Sinclair Sequences, sediments of the Neoproterozoic to Cambrian Nama Group and nappes of the Naukluft Nappe Complex. The basement is found mainly on the western and northern side of the mountains, and consists of meta-sedimentary and meta-volcanic rocks, gneisses and granites, varying in age between 1800 and 1000 Ma. Sediments of the Kuibis and lower Schwarzrand Subgroups of the Nama Group overlie the basement with a major regional unconformity. The top of this unit is truncated and overlain by four intensely folded and imbricated allochtonous nappes, containing several lithologically distinct sequences of Neoproterozoic age. Limestone, dolomite and shale form the Dassie Nappe, while the Zebra Nappe is composed of quartzite, black limestone, conglomerate, shale and dolomite. Shale, conglomerate and dolomite make up the Pavian Nappe, and the Kudu Nappe contains dolomite, limestone, shale and conglomerate. This unit of nappes is separated from a single lower nappe by a thin dolomite band, the Sole Dolomite or so-called "Unconformity Dolomite" of Korn & Martin (1959) (Fig. 8.22.1).

The Naukluft nappes, which form the top part of the mountains, are very large, sheet-like bodies, which were emplaced along low-angle thrusts, with movement towards the S and SE. Comparative lithological studies relate the rocks to the Swakop Group of the Damara Sequence and place the source of the nappes in the region of the Hakos Mountains to the N, a minimum displacement of some 78 km. The movement occurred some 500 to 550 Ma ago related to the Damaran Orogeny. Earlier nappes became imbricated during deformation and moved as a single unit on the Sole Dolomite into their present position. They overrode the underlying Nama rocks, imbricating them to form a single parautochthonous nappe, the Rietoog Nappe, below the Sole Dolomite.

The Sole Dolomite played an important role in the process. Before crystallization, it was a mobilized, mylonitic sludge with a high content of hot, saline fluids. These fluids were probably derived from evaporite beds in the lower Damaran Duruchaus Formation, which today fills a deep basin some 90 km to the northeast of the Naukluft, but may originally have had a wider extent. The saline fluids reduced the frictional strength at the base of the upper nappe complex, and as the angle of southward tilt increased, the whole complex could slide southeastwards with the fluids at its base as a lubricant (Martin et al. 1983).

Younger dissolution of the dolomites and limestones of the Naukluft Complex has given rise to karstification of the plateau with a resulting extensive underground drainage system. In some of the deeply incised valleys discharge from this underground water occurs in the form of crystal clear fountains, and tufa, also known as fountain stone, occurs associated with these springs. A large tufa formation can be seen from the road on farm Blässkranz, and actually gave the farm its name (Fig. 8.22.2). The

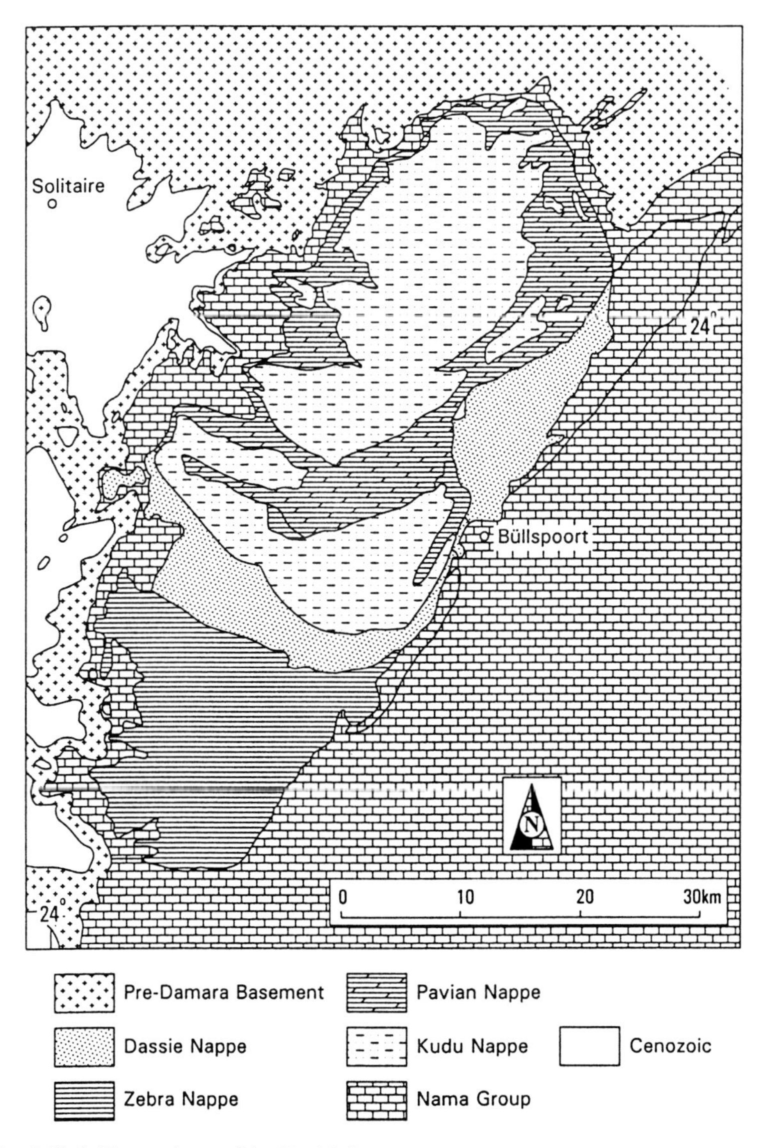


**Fig. 8.22.1:** The geology of the Naukluft.

**Fig. 8.22.2:** The tufa formation on farm Blässkranz.

soft, semi-friable, porous limestone is formed by evaporation of the calcium carbonate-rich water. Two particularly impressive tufa formations can also be seen on the Waterkloof Trail.

## 8.23 The Omatako Mountains

About 85 km N of Okahandja, just to the W of the main road between Okahandja and Otjiwarongo, are two conical inselbergs, which rise between

700 and 800 m above the surrounding plain. They are called the Omatako Mountains, Omatako being the Herero word for buttocks, and the larger one of them at a height of 2289 m above sea level is the second highest mountain in Namibia north of the Erongo. The other attains a hight of 2175 m above sea level.

The Omatako Mountains are erosional relics of the sedimentary and volcanic rocks of the Karoo Sequence, which were deposited between 180 and 220 Ma ago in an area extending to Mount Etjo and the Waterberg in the N. The lower portion of the western peak is composed of sandstone, shale, mudstone and conglomerate of the Omingonde Formation. Halfway up the mountain the Omingonde Formation is terminated by a 80 to 90 m thick layer of aeolian sandstone of the Etjo Formation, which in turn is capped by 300 m of basalt. The eastern cone consists of the same sedimentary units as the western one, however, in this case the capping consists of dolerite, the coarser grained intrusive equivalent of basalt. This dolerite capping is the continuation of a 150 m wide dyke, and which is clearly visible on the northern and southern slopes of the mountain (Fig. 8.23.1).

Sediments of the Omingonde Formation were deposited in lakes and by rivers about 220 Ma ago during semi-arid climatic conditions. The slopes of the mountains are covered by boulders of the overlying basalt and do-

**Fig. 8.23.1:** The N Omatako Mountain.

lerite and somewhat conceal the sediments of the Omingonde Formation, nevertheless, the sandstone of the Etjo Formation forms a steep cliff, and can easily be recognized. The deposition of aeolian sandstone marks a significant change in climate to much more arid conditions about 190 Ma ago. Finally, the volcanic activity, which produced the basalt and the dolerite, was part of the wide-spread Karoo volcanism, which is related to the early break-up of Gondwanaland. Erosion has since removed the vast majority of the abundant volcanic rocks with the capping of the Omatako Mountains proving to be more resistant and so preserving a section through the underlying Omingonde and Etjo Formations.

## 8.24 Organ Pipes

The organ pipes, 6 km E of Twyfelfontein, are located on a short detour en route to Burnt Mountain. About 3 km after the Twyfelfontein turn-off on road 3254, and some 7 km before Burnt Mountain is reached, a small road turns of towards the E and leads to the Organ Pipes. At this locality, a do-

**Fig. 8.24.1:** The Organ Pipes.

lerite sill, related to the break-up of Gondwanaland in the Early Cretaceous, has intruded Damaran metasediments and Karoo sediments. As the dolerite cooled, it developed a system of intersecting, contractional fractures termed "columnar jointing", which upon weathering, give the appearance of organ pipes, from which the locality gets its name.

Columnar jointing consists of a close-packed series of hexagonal, sometimes pentagonal or heptagonal, prisms lying perpendicular to the upper and lower surfaces of a lava flow or sill. The joints are formed during the cooling of the lava by contraction and by the resultant tensional forces acting towards a number of centers within one layer. These forces tend to pull open a series of joints, which ideally assume a hexagonal pattern (Fig. 8.24.1), and, as cooling proceeds towards the central part of the mass, the joints develop at depth (Whitten & Brooks 1981).

## 8.25 Dinosaur Footprints at Otjihaenamaparero

Dinosaur footprints occur on the farm Otjihaenamaparero 92 in the Otjiwarongo District, some 64 km NW of the turnoff from the Okahandja – Otjiwarongo main road of secondary roads D2404 and D2414. The farm is sign-posted.

The tracks occur in sandstones of the 200 Ma old Etjo Formation. The sands forming these sandstones accumulated under increasingly arid conditions as wind blown dunes similar to those existing today in the Namib Desert. Numerous reptiles lived in the interdune areas, but as the climate became drier, these animals were forced to concentrate near waterholes, small lakes and rivers fed by occasional rainfalls and thunderstorms. Inevitably, their feet left imprints in the wet sediment around the water. Later, these imprints were covered by other layers of wind blown sand, and were preserved as trace fossils when the sand solidified into rock due to the pressure that built up as they became buried ever deeper.

At Otjihaenamaparero, two crossing tracks consist of more than 30 imprints with a size of approximately 45 by 35 cm. The longer track can be followed for about 28 meters (Fig. 8.25.1). There is a distance of some 70 to 90 cm between individual imprints. In addition, there are a number of individual imprints as well as some tracks comprising smaller imprints of about 7 cm length and a path length of 28 to 33 cm (Gührich 1926). All

**Fig. 8.25.1:** Dinosaur footprints on farm Otjihaenamaparero.

tracks clearly show the form of a three-toed, clawed foot, and from their arrangement it is deduced that they were made by the hind feet of a bipedal animal. Unfortunately, no body fossils have been found in the area so far, and one can therefore only use comparison with other sites for identification.

Although worldwide, about 900 dinosaur species are known through the discovery of body fossils, only a few dozen footprint types have been discovered and identified (Lockley 1991). From these it can be concluded that the Otjihaenamaparero dinosaur possibly belonged to the large carnivore order of Therapoda. The dimensions and the depth of the imprints suggest

**Fig. 8.25.2:** Reconstruction of the Otjihaenamaparero dinosaur (after Lockley 1991).

that the dinosaur had an appreciable size, and might have looked like the reconstruction in Fig. 8.25.2. Due to the unfavourable changes in climate, it can be assumed that the animals became extinct not long after they left their footprints.

There are a number of localities in the Etjo Sandstone that contain dinosaur footprints, however, Otjihaenamaparero is the most impressive. The site has been declared a National Monument, and the footprints are protected by law.

## 8.26 Petrified Forest

Although the occurrence of petrified wood in rocks of the lower Karoo Sequence is not uncommon, the "Petrified Forest" 45 km W of Khorixas is the biggest accumulation of large petrified logs in southern Africa. The logs are in an excellent state of preservation and the Petrified Forest is a declared National Monument, and no samples may be taken. The logs occur at the 280 Ma old base of the Permian Ecca Group of the Karoo Sequence and have been deposited in an ancient river channel. The matrix carrying the logs is a brownish, cross-bedded sandstone (Fig. 8.26.1).

Recent erosion has exposed many of the logs and also several smaller pieces. The larger logs are up to 1,2 m in diameter and at least two trees are exposed with a full length of 45 m. Although the trunks are broken into sections 2 m and shorter, the individual segments are still in place. The trunks are straight and taper gradually. Several hundred different logs are partly or completely exposed, appearing as if they had drifted into their present position in the old river sediments. Such a drift, for example during a heavy flood event, and the associated rapid deposition and burial, provides a good explanation for the concentration of large amounts and good state of preservation of the fossilized wood.

The petrified wood pieces belong to seven different species of the collective type *Dadoxylon arberi* Seward (Kräusel 1928, 1956). The wood has

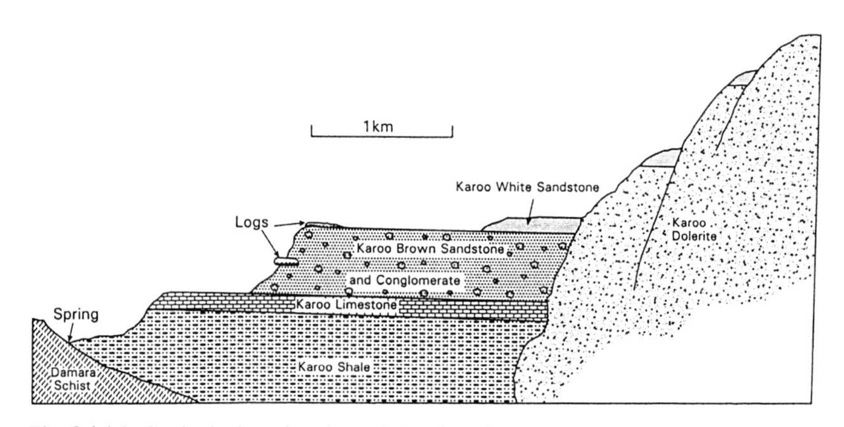


**Fig. 8.26.1:** Geological section through the site of the Petrified Forest.

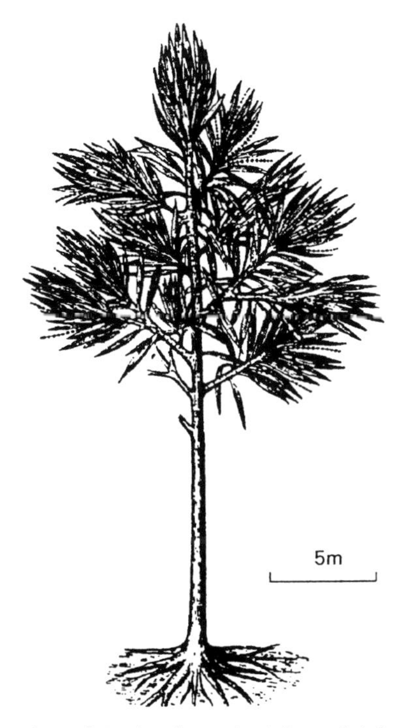


**Fig. 8.26.2:** Reconstruction of *Dadoxylon arberi* Seward (after Stewart & Rothwell 1983).

been silicified and agatized except for some parts, which are filled with calcite. The colour varies from brown with white streaks to red with light coloured streaks, and some pieces are also white. Presence of annual growth rings of varying thicknesses indicates that the trees grew in a seasonal climate with pronounced rainfall variation. Cell structures are well preserved.

*Dadoxylon arberi* Seward was a conifer belonging to the now extinct order Cordaitales of the Gymnospermopsida class (Fig. 8.26.2). The woody plant formed a highly branched tree with simple, needle-like leaves. The simple pollen cones were not more than a few centimeters in length. The root system is believed to have been shallow and extending laterally for several meters (Stewart & Rothwell 1983).

## 8.27 Sesriem Canyon

by John Ward

The Sesriem Canyon is a narrow, 30 m deep canyon on the Tsauchab River, which was caused by the incision of this drainage into, and through, calcified gravels of its ancient course on the eastern edge of the Namib Desert (Fig. 8.27.1). This down-cutting, which probably occurred during the Late Miocene some 5 Ma ago, has exposed the dolomite- and limestone-rich gravels of the Karpfenkliff Formation, deposited on a former braidplain of a Proto-Tsauchab River system during the Middle Miocene. The clasts of these gravels were derived from Neoproterozoic carbonates of the Naukluft Mountains and the Kuibis and Schwarzrand Subgroups of the Nama Group in the catchment area of the Tsauchab River on the Great Escarpment to the E. In places at the bottom of the Sesriem canyon, older sandstones of the Tsondab Sandstone Formation of the Proto-Namib desert phase have been exposed by the fluvial erosion processes that continue to carve this small but striking feature.

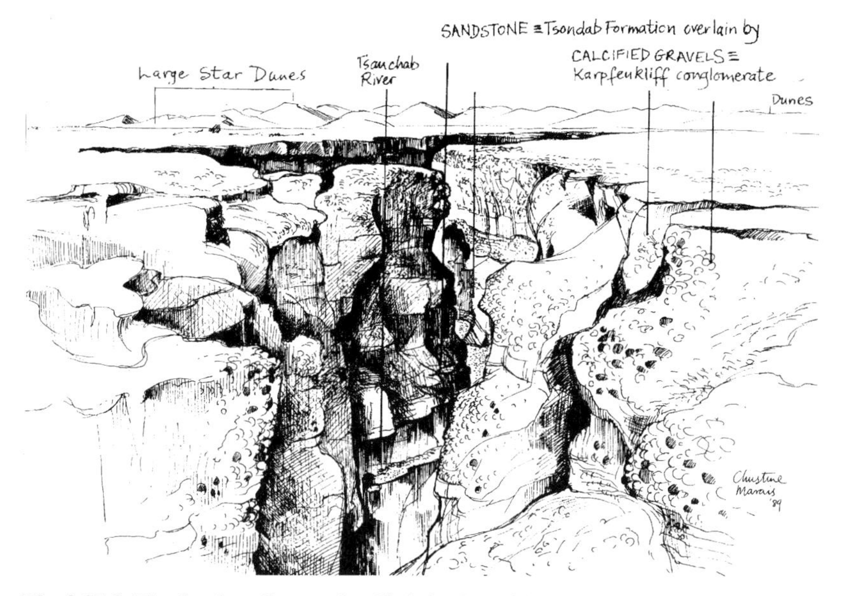


**Fig. 8.27.1:** The Sesriem Canyon (by Christine Marais).

The name Sesriem is derived from the fact that in the past, the early settlers in this dry area had to attach six ("ses") length of oryx hide straps ("riem"), normally used for their ox wagons to harness the oxen, to reach the water in the deep canyon with a bucket. Individual straps varied in length between 3 and 6 meters.

## 8.28 Sossusvlei and Tsondabvlei

by John Ward

The Tsauchab and Tsondab Rivers are comparatively small, ephemeral drainages that rise S and E, as well as out of the Naukluft Mountains and cut westwards across the Namib Desert to end in terminal pans within the main Namib Sand Sea. In recent geological history, the northward migrating dunes of the main Namib Sand Sea have blocked the middle reaches of both rivers forming the impressive pans of Sossusvlei on the Tsauchab River and of Tsondabvlei on the Tsondab river. The endpoint or terminal pans are composed largely of pale coloured calcareous silts derived from the carbonates situated in the Naukluft and Zaris Mountains of the Great Escarpment. Significantly, gravels characteristic of both catchments are exposed amongst the dunes to the W of both Sossusvlei and Tsondabvlei, indicating earlier extensions westwards to the Atlantic Ocean, possibly at Fischersbrunn south of Meob Bay and at Conception Bay respectively. In places, the silts of these former extensions harbour fossilised shells of freshwater gastropods.

Run-off down the Tsauchab and Tsondab Rivers is dependent on summer rainfall in the respective catchments – an extremely variable factor in this marginal zone east of the Namib Desert. Thus, it is only in years of exceptional summer rainfall, *inter alia* 1963, 1974, 1976, 1987, 1997, 1999 and 2000 that the Tsauchab River flowed through to Sossusvlei and likewise the Tsondab River pushed through to its endpoint (Fig. 8.28.1). Depending on the run-off, and hence the volume of water reaching Sossusvlei and Tsondabvlei, these endpoint pans may hold water for several months. Once dry, the pans display typical mud cracks, which are irregular fractures in a crudely polygonal pattern, formed by the shrinkage of clays in the drying process.

**Fig. 8.28.1:** Water in the desert during the 1997 flood of the Tsauchab River.

**Fig. 8.28.2:** Dead tree at Dead Pan S of Sossusvlei.

Remnants of former pans can be found to the S, as well as to the W of Sossusvlei. Dead trees in one of the southern pans gave a radiocarbon date of about 900 years before present (Fig. 8.28.2). The highest, free-standing star dune in the Sossusvlei area rises some 200 m above the floor of the ancient pan. On the northern side, however, the prominent reversing arm of the star dune rises some 325 m above the valley floor. These heights are attained, because the star dunes are resting on an elevated terrace cut into Tsondab Sandstone, which forms the northern side of the Tsauchab Valley. They have, nevertheless, contributed to the common notion, that the dunes in the Sossusvlei area are the highest dunes in the World.

## 8.29 Spitzkuppe

by Steven Frindt and John Kandara

The Grosse and Kleine Spitzkuppe are Cretaceous anorogenic complexes, located in central western Namibia, some 40 km NW of Usakos and N of the road to Henties Bay. The Grosse Spitzkuppe, also known as the Matterhorn of Africa, is a declared National Monument. It has a height of 1728 m above sea level. The complex has a diameter of some 6 km and towers more than 670 meters above the Namib plain, providing a challenge to many a rock climber. The Pontok Mountains to the immediate E also belong to the complex and attain a height of 1628 m above sea level.

The neighbouring Kleine Spitzkuppe, situated some 15 km W of the Grosse Spitzkuppe, has a diameter of slightly more than 6 km and reaches a height of 640 meters above the surrounding peneplane, with a total height of 1584 m above sea level. However, most of the granite of Kleine Spitzkuppe occurs as low-lying micro-granite outcrops.

The Spitzkuppe granites have interesting erosional formations, sculpted by the persistent W winds and extreme differences between night and day temperatures. During the cooling of the granite magma, numerous joints formed due to the decrease in volume during the transition from the liquid to the solid phase. Between these joints, the granite is arranged in concentric layers according to the progressive cooling from the inside to the outside. Erosion later removed layer by layer, leaving the typical rounded granite boulders. This type of weathering is termed "onion-skin weathering" and typical of granitic rocks in subtropical regions.

The emplacement of the Spitzkuppe granites in metamorphic rocks of the Damara Sequence was part of the widespread magmatic activity in Namibia which occurred with the opening of the South Atlantic Ocean during Gondwana break-up some 130 Ma ago. Erosion of overlying country rocks led to the exposure of the Grosse and Kleine Spitzkuppe granite complexes, with the roof pendant now completely eroded away. Locally the country rocks comprise basement rocks of Damaran granite and metasediments of the Damara Orogen. The Spitzkuppe granites were preceded by a period of basaltic-rhyolitic magmatism, evident as mainly N-NE trending dyke swarms. The emplacement of the Spitzkuppe granites into the high crust is probably due to cauldron subsidence.

The Grosse Spitzkuppe is a zoned granite complex (Fig. 8.29.1) comprising of three texturally distinct granite phases, a marginal, medium-grained biotite granite, a coarse-grained biotite granite and a central porphyritic granite. Late stage aplite and porphyry dykes, as well as lamprophyre dykes cut the granite phases of the complex. Syn-plutonic mafic dykes and mafic to intermediate magmatic enclaves within the granites demonstrate the bimodal character of magmatism. The outer contacts of the complex and the upper margins of aplite dykes are marked by banded pegmatite-aplites and fan-shaped alkali feldspar growth, which resulted from under-cooling of the magma. The presence of miarolitic cavities and pegmatite pockets are common, especially in the outer parts of the medium-grained biotite granite, in the porphyritic granite and in aplite dykes, indicative of fluid saturation during the final stages of crystallization.

The granite phases contain quartz (31 – 38 %), alkali feldspar (30 – 42 %) and sodic plagioclase (16 – 25 %). Biotite typically represents less than 6 % of the rock volume. Magmatic topaz is an ubiquitous minor constituent occurring as subhedral grains. Other common accessory minerals include tourmaline, beryl, fluorite, columbite, monazite, thorite, magnetite, and niobian rutile.

The Kleine Spitzkuppe complex consists of two main granite types (Fig. 8.29.1), an equigranular medium-grained granite that forms the Kleine Spitzkuppe Mountain, and numerous low-lying outcrops of equigranular to porphyritic microgranite containing abundant miarolitic cavities and pegmatite pockets. Both granite phases are composed of quartz, alkali feldspar and sodic plagioclase, with biotite and topaz occurring as minor constituents, and columbite and beryl as accessory minerals.

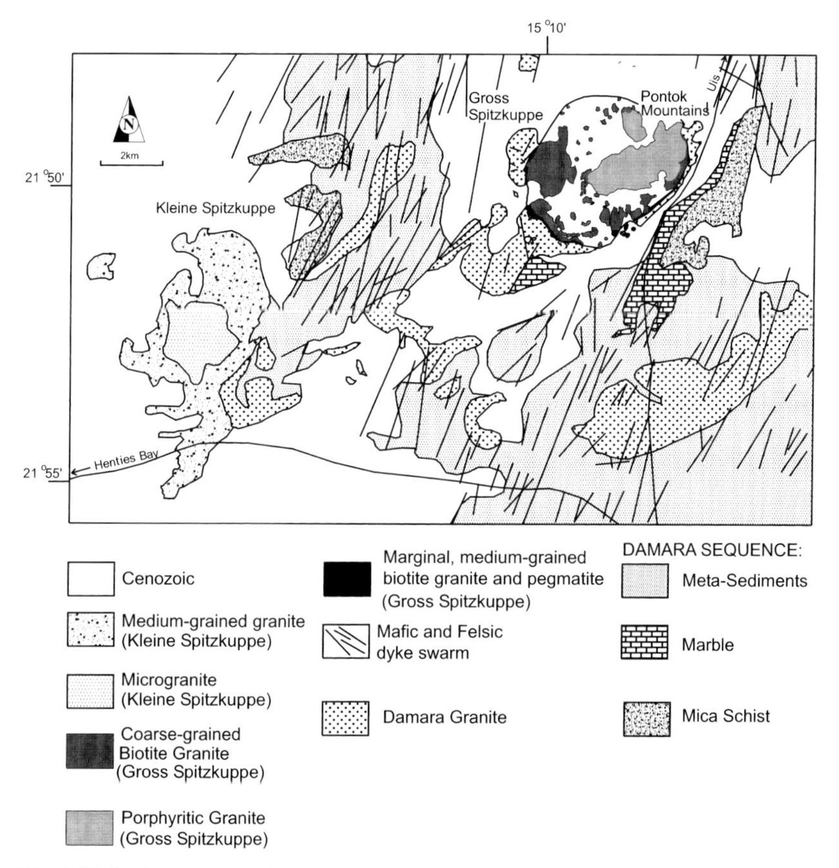


**Fig. 8.29.1:** The geology of Grosse and Kleine Spitzkuppe.

Minor intrusive phases include a number of flow-banded rhyolitic dykes that cut into the country rock and a few, approximately 1 m wide lamprophyre dykes occurring within the complex. The contact with the Damaran country rocks is often marked by layered marginal pegmatite from which local inhabitants mine gem quality topaz and aquamarine crystals. The "African Granite" company operates a quarry for dimension stone south of the Kleine Spitzkuppe.

Geochemically, the granites of the Grosse and Kleine Spitzkuppe are very similar, with little compositional variation between the granite phases of each complex. According to Streckeisen's classification of plutonic rocks, the Spitzkuppe granites can be classified as monzogranites. These granites are high in silica, subalkaline and marginally peraluminous, and are characterized by higher Si, Fe/Mg, F, Zn, Ga, Rb, Zr, Hf, Th, U, and REE (except Eu), and by lower Ca, Mg, Al, P, B, and Sr when compared to average granitic rocks. The granites of the Grosse and Kleine Spitzkuppe are interpreted to have formed by crustal re-melting related to mafic underplating in a continental rift environment.

## 8.30 Twyfelfontein

by John Kinahan and Gabi Schneider

Twyfelfontein lies in a valley running northwards and carrying a small tributary of the Huab River. The valley is bounded by sandstones of the Etjo Formation and shales of the Gai-As Formation, Karoo Sequence, that are underlain by dark Kuiseb Formation schists of the Damara Sequence. The geological juxtaposition of porous aeolian and fluvial sandstones on impermeable deposits resulted in the formation of a freshwater spring. The name "Twyfelfontein", meaning "doubtful spring", originated because it only carries water episodically, the porous aeolian sandstone being limited in volume and only holding a limited amount of water after good rainy seasons. Undoubtedly, people have for a long time been attracted to the area by this small spring which brings a great variety of game to the area, which hunters can observe unseen from a terrace some 50 m above the spring itself.

The rocks of the Etjo Formation, a thick bedded aeolian sandstone, weather into large blocks, often with clean flat faces. These large blocks provided great shelter and the large flat faces, the old dune slip faces, the "canvas" for rock art. In 1952, this rock art at Twyfelfontein was proclaimed a National Monument.

With almost 2000 recorded images, Twyfelfontein or Ui-ais, as it was originally called, is the largest known rock engraving site in Namibia, and one of the largest in Africa. Most of the engravings, as well as a small number of paintings, are well preserved and their subject matter is easily iden-

**Fig. 8.30.1:** Rock engravings on Etjo sandstone at Twyfelfontein.

tifiable (Fig. 8.30.1). The most prominent species are giraffe (234) and rhino (121), followed by zebra (75), oryx (50), ostrich (40), elephant (24), and cattle (29). In contrast to the rock painting sites in the same area, this and other engraving sites have virtually no human figures. However, they include many more birds than the painting sites and it is interesting to observe that most of the birds are the so-called striding species, such as the ostrich, rather than perching species, such as sparrows. This suggests that the birds may in fact represent people. There are many other indications that the engravings are not simple naturalistic representations, and the clearest, but most easily overlooked example, is the engraving of a lion which has five toes, like a human, rather than the usual four.

At Twyfelfontein, the rock engravings occur in eight small clusters, on the high terrace and in the small ravine to the west. The whole area is littered with stone artefact debris, mainly vein quartz and indurated shale, which would have been carried from far afield. The small spring at this site probably served as an essential water supply for thousands of years, with hunter-gatherer communities eventually giving way to livestock farmers. In years of plentiful rain, the area to the west would have been used until the onset of the dry season, and smaller rock engraving sites are found as far away as the edge of the Namib Desert. The distribution pattern of these sites indicates a network of alternative hunting grounds and the routes which connected them. Interestingly, the archaeological evidence shows that the animals most commonly hunted were small antelope, hyrax and even lizards, rather than the large species depicted in the engravings which apparently had more ritual than subsistence value.

## 8.31 Vingerklip

Vingerklip is a prominent, 30 m-high erosional remnant on the farm Bertram 80 in the Ugab River Valley SW of Outjo. The Ugab River, from its source in eastern Damaraland, follows an almost direct W-SW course to the sea and has incised a broad valley into the Early Tertiary landsurface. The Ugab has a steep gradient and during periodical floods incises itself into older terraces. Three terrace levels are present in the valley, with the main terrace at 160 m above the present river (Fig. 8.31.1). It rests on an Eocene surface and contains more than 100 m of sand and sandy, calcrete-cemented conglomerate, capped by calcareous sandstone. Late Tertiary uplift resulted in a mainly erosional Lower to Middle Pleistocene terrace 30 m above the present river, followed by an Upper Pleistocene wet phase, which caused erosion below the present floodplain. Subsequently deposited sandy alluvium containing Middle Stone Age tools is at present undergoing dissection into a lowermost terrace (Mabbutt 1950).

Vingerklip is a remnant of the main terrace which is best preserved on the northern side of the valley. It consists of calcrete-cemented conglomerate and calcareous sandstone of the main terrace. Boulders in the conglomerate have been derived from Damaran schists, marbles and quartzites and pre-Damaran gneisses in the Ugab River catchment area. The intercalations of conglomerate and sandstone give an indication of the changing sedimentologial conditions in the river during the last 40 Ma. The hardness of the conglomerate and its capping, which facilitated the survival of this erosional remnant, is due to the proximity of the neighbouring dolomite

**Fig. 8.31.1:** The Ugab Terraces, Inset: Vingerklip.

ridges in the N, which yielded the carbonate matrix of the terrace capping. The less indurated valley fill on the southern side of the Ugab River has succumbed more easily to erosion. From the site of Vingerklip one has a magnificaent view of the main terrace of the Ugab River.

## 8.32 The Waterberg

The Waterberg of NE Namibia is situated some 60 km E of Otjiwarongo and forms a table mountain with a prominent plateau of some 40 000 hectares. The highest point lies at 1930 m above sea level. To the NW, the SW and the SE, the mountain is surrounded by steeply rising slopes. The plateau dips gently to the NE where the mountain eventually merges with the Kalahari Sandveld to the E.

The geological history of the Waterberg began 300 Ma ago, when thin tillites were deposited on Damaran basement by the retreating Dwyka ice sheet (Fig. 8.32.1). Thereafter, shales, mudstones and sandstones of the Ecca Group were deposited, however, only locally. About 220 Ma ago, a half-graben, the Omingonde Basin, developed in the Waterberg and Mount Etjo (see 8.19) region and sedimentation of the Omingonde Formation started in this half-graben under semi-arid climatic conditions. Up to

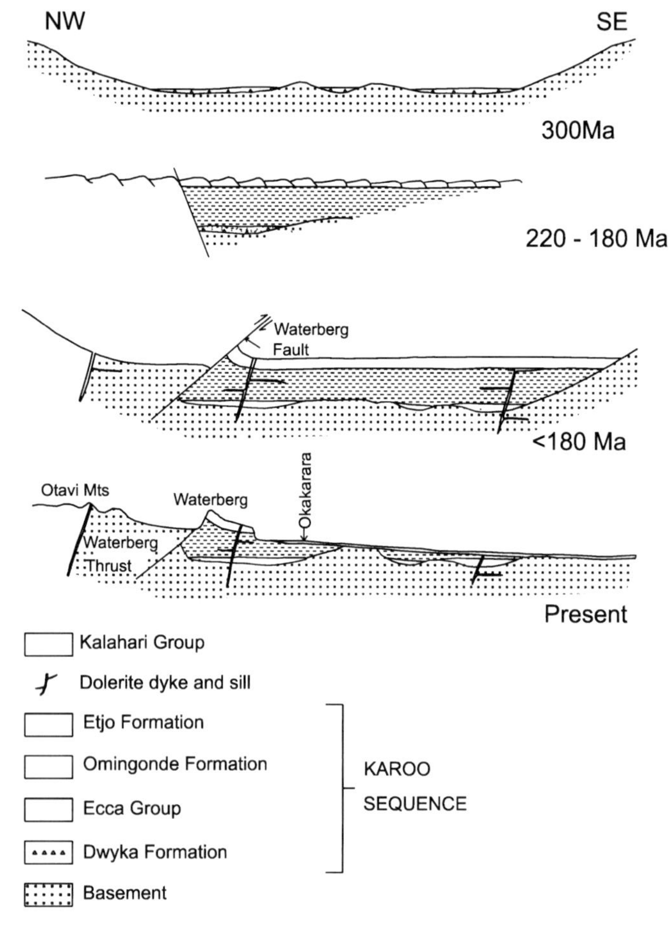


**Fig. 8.32.1:** Geological evolution of the Waterberg.

700 m of conglomerates, sandstones and mudstones accumulated, and some limestone layers also occur. These sediments were derived from the neighbouring highlands, where the basement rocks were exposed to weathering. Rivers transported the eroded material into the lakes in the half-graben, however, they probably flowed only intermittently after periodic rains, as is evident from the re-working within individual sediment layers. As time progressed, the climate became more arid and about 180 Ma ago, dunes covered the entire area, which now lithified, form the Etjo Formation overlying the Omingonde Formation. At the Waterberg itself, the Dwyka and Ecca Groups do not crop out. The Omingonde Formation consists mainly of massive, red-brown, fine-grained sandstone and siltstone with subordinate mudstone.

The Waterberg Fault is a prominent feature in northern Namibia and runs for about 250 km from Omaruru to Grootfontein and past the northwestern foot of the Waterberg. Along this fault the basement and Karoo rocks were thrusted upwards and eventually over the Karoo rocks, protecting them from weathering, and thereby playing a decisive role in the formation of the Waterberg (Fig. 8.32.1). The vertical displacement of the rocks adjacent to the thrust is at least 700 m. Associated with Gondwana break-up some 130 Ma ago, rapid continental uplift led to the formation of the central depression of the Kalahari Basin. The Waterberg is situated on the western side of the Kalahari Basin, where the strata were tilted to the NE in the process. Along the northwestern edge of the plateau the Okarukuvisa Mountains, which were formed by upward tilting of the Karoo rocks as the basement was thrown up against them, protect the mountain from erosion. Today, the mountain stands as an erosional relic with the harder Etjo Sandstones protecting the underlying Omingonde Formation from erosion.

Modern weathering of the Omingonde Formation undercuts the more resistant Etjo Sandstone. Fragments of the sandstone break from the edge of the plateau and cover most of the slopes in scree and boulders, as can be easily seen during the climb from the restcamp to the plateau. The process is intensified by springs. Rainwater percolates down into the porous aeolian sandstones of the Etjo Formation until it encounters the impervious mudstone or siltstone of the Omingonde Formation. Due to the inclination of the strata the water re-appears in springs along joints near the boundary of the Omingonde and Etjo Formations, giving the mountain its name. Intensified weathering near the springs has resulted in the formation of steep, narrow cliffs. Where the slope is covered by scree and boulders, water seepage results in swampy areas with lush vegetation, as for example near the restcamp.

Although the depositional environment of the Omingonde and Etjo Formations was not very favourable for the preservation of fossils, some were found, including dinosaur tracks. The tracks were preserved high up in the Etjo Formation, at least 100 m above its base. The size and form of the tracks is similar to the larger ones at Otjihaenamaparero (see 8.25), and, like the dinosaurs that left their footprints there, the dinosaurs of Waterberg probably concentrated around shallow pools in inter-dune areas in the otherwise arid environment. Also, parts of the skeleton of one of the oldest known dinosaurs, *Massospondylus*, were found in sandstones in the lower Etjo Formation at Waterberg. *Massospondylus* lived 200 Ma ago and was a prosauropod herbivore of about 3 to 5 m length and 1,5 m hight. It walked on four legs and was a relatively fast runner, probably on the two hind legs (see 5 and 8.25).

# 9. Excursions

The following section of the Roadside Geology Guide provides descriptions of routes, which were selected taking into consideration the major attractions that a visitor to Namibia, as well as the local traveler, is most likely to visit. They are presented in a radial fashion, starting with routes from the central part of the country towards the W, followed by S and E and ending with the northern routes (Fig. 9.1).

Namibia is fortunate to have one of the best road infrastructures on the African continent, and with the exception of route 9.29, which is strictly 4 x 4 and requires off-road skills, all routes can be negotiated with a 2 x 4 sedan car. It is, nevertheless, not recommended that sedan cars travel routes 9.3 and 9.5 in the direction from W to E, because of the steep gradient of the passes that need to be negotiated. While the major roads in Namibia are surfaced, there are many regional connections, which are gravel roads. These are usually of excellent quality, and are regularly maintained. This often gives the wrong impression that one can travel at high speed as many are almost like a surfaced road. This is not recommended, and speeds should not exceed 100 km/hour, likewise over-steering needs to be carefully avoided. Last, but not least, wild animals frequently cross the roads, and should be watched out for, particularly at dusk or at night.

Large parts of Namibia are Nature Conservation areas and some of these require entrance permits, whilst others do not. Route 9.4 crosses the Namib Naukluft Park, but a permit is only required to visit the lookout point into the Kuiseb Canyon. Routes 9.5 and 9.6 also cross the Namib Naukluft Park, but no permit is required. Route 9.24 ends in the Etosha National Park, for which a permit is required. Route 9.27 enters the Skeleton Coast Park, and likewise a permit is required. All these permits can be obtained from Namibia Wildlife Resorts, Private Bag 13378, Windhoek, Tel +264-61-25 64 43, Fax +264-61-25 67 15 or nwr@mweb.com.na. In the case of route 9.10 which enters Diamond Area No. 1, a permit must be obtained through the Security Division of Namdeb Diamond Corporation, PO Box 35, Oranjemund. However, permits to travel to Oranjemund are at present only issued under certain circumstances, and not freely available to the tourist. A police clearance certificate is required for the permit application. All other routes can be traveled without permits.

A large variety of accommodation is available in Namibia. There are five-star hotels, guest farms, pensions, bed-and breakfast establishments, municipal bungalows and camp sites, as well as camps in the national parks, which provide bungalows as well as camp sites. The camps in the national parks can be booked through Namibia Wildlife Resorts, at the address given above. A guidebook giving addresses for other accommodation is published annually by the Namibia Tourism Board, Private Bag 13346, Wind-

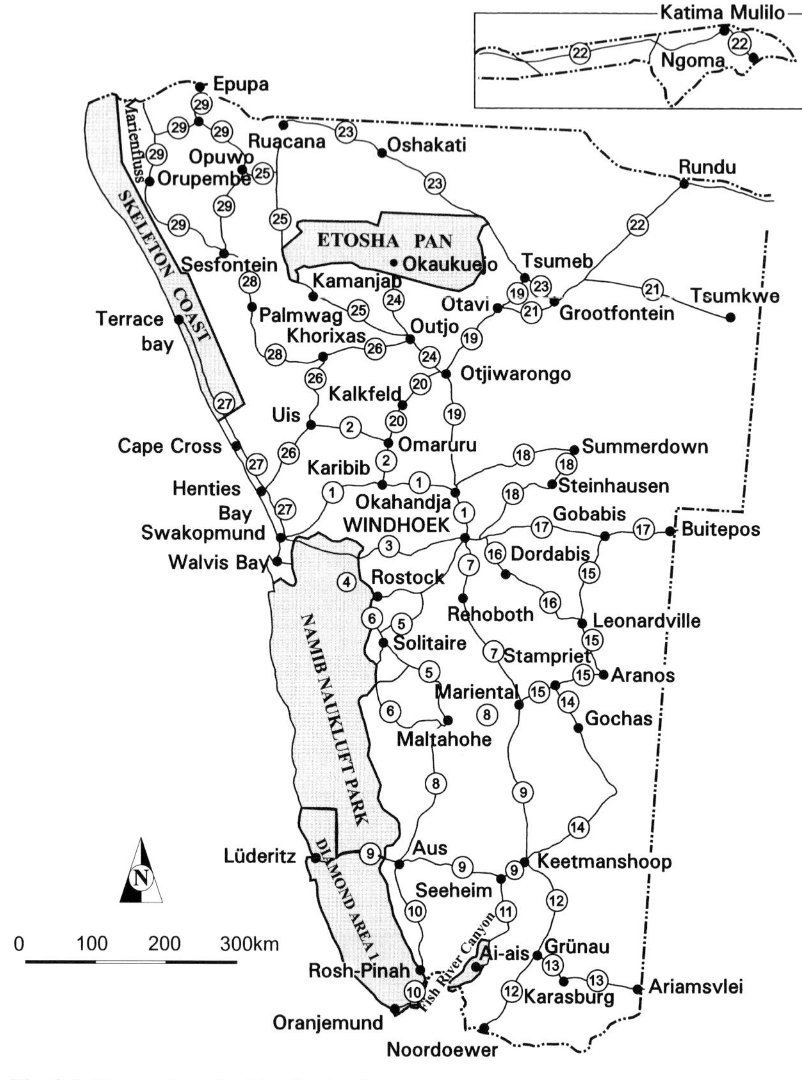


**Fig. 9.1:** Routes described in chapter 9.

hoek, Tel/Fax +264-61-28 42 360 or tourism@mweb.com.na. The Ministry of Environment and Tourism also has information offices in a number of countries.

The following route descriptions are accompanied by geological maps as well as stratigraphic diagrams showing the stratigraphic position of the rocks encountered en route. The time scale on the right side of the diagrams is provided to give an approximate correlation of the strata with specific rock-forming periods in the Namibian geological time scale as given in section 3. It does not mean, however, that the formation of the rocks always lasted for that entire period.

Topographic maps of Namibia are available at scales of 1:1 million, 1:250 000, 1:100 000 and 1:50 000 from the Office of the Surveyor General, Private Bag 13182, Windhoek, Tel +264-61-50 56, Fax +264-61-24 98 02. A large variety of geological maps at various scales, including thematic maps, and geological reports can be obtained from the Geological Survey of Namibia, Private Bag 13297, Windhoek, Tel +264-61-2 08 51 11, Fax +264-61-24 91 44 or secretary@mme.gov.na. The black and white maps provided in this guide were taken from the coloured 1:2 million scale Geological Map of Namibia, which is also available from the Geological Survey of Namibia.

## 9.1 Windhoek – Okahandja – Karibib – Swakopmund

The first part of this route follows road B1 from Windhoek to Okahandja. The area between these two towns belongs to the Southern Zone of the Damara Orogen and is underlain entirely by schist of the Kuiseb Formation (Figs 9.1.1, 9.1.2). This Southern Zone developed from an intra-continental rift into a narrow ocean, where the widespread greywacke succession of the Kuiseb Formation, now metamorphosed into schist, was deposited. The schist forms a highly thrusted, kyanite-bearing, high-pressure, low-temperature belt with a southward vergence (Miller 1983,1990).

The road follows the Windhoek Graben, which is part of a prominent fault system of Mid-Tertiary age developed as a result of extensional tectonics. Such extensional tectonics have occurred in the Earth's crust in southern Africa since the Early Cretaceous some 130 Ma ago, heralding the break-up of Gondwanaland. Activities on these faults are still, albeit to a minor extent, indicative of extensional tectonics continuing as the continents drift further apart.

The Windhoek Graben is about 150 km long and 20 km wide. To accommodate crustal stress, individual blocks were displaced vertically in relation to each other, along steeply dipping faults, with only minor horizontal movement. Schist forms the prominent mountain range to the E of the road, and also the hills in the W, and for about 30 km they can be seen in numerous outcrops close to the road and in road cuttings, but thereafter the Windhoek Graben widens to the N, and outcrops become rare. Also of interest are Cenozoic dunes, which have developed on the floor of the Windhoek Graben, and can be seen in road cuts about 25 km from Windhoek.

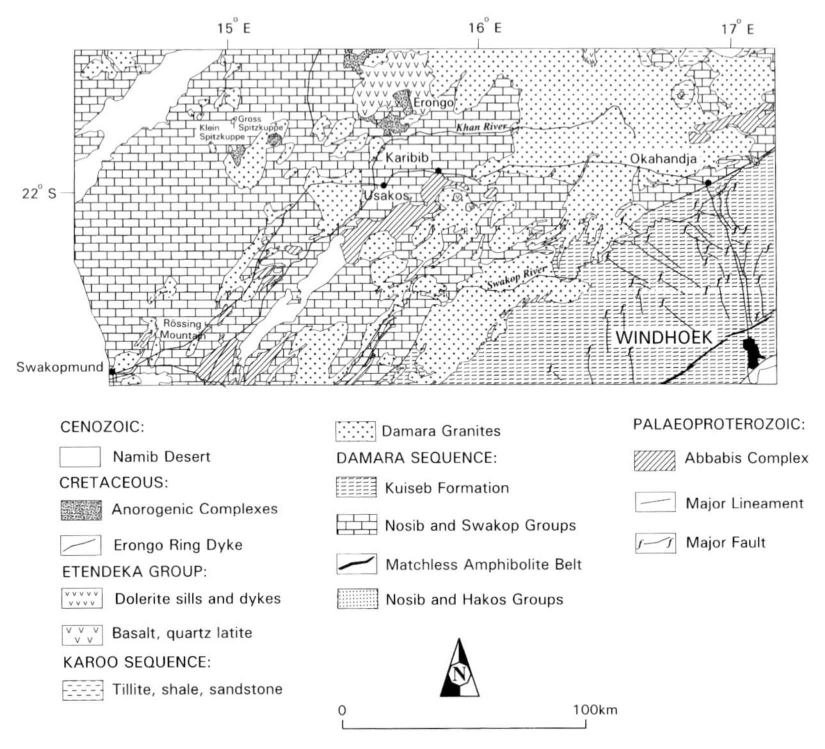


**Fig. 9.1.1:** Geological map for route 9.1.

Before Okahandja is reached, two bridges cross the Swakop River, one of Namibia's major ephemeral rivers, and its tributary, the Okahandja River. The waters of the Swakop River are collected upstream in the Von Bach Dam, a major source of water for Windhoek, and the river therefore only floods after exceptional rains in the catchment area downstream from the dam. Between the rivers, on the southern outskirts of the town, outcrops of quartzose and pelitic mica schist of the Kuiseb Formation intruded by pegmatites occur next to the eastern side of the road. Porphyroblastic cordierite up to 2 cm long is present in the pelitic layers.

It is interesting to note, though not readily recognized, that just S of Okahandja the Okahandja Lineament, one of the most important boundaries in the Damara Orogen, is crossed. It separates the Central Zone of the Damara Orogen from the Southern Zone and marks major changes in stratigraphic succession, structural style and age of tecton-

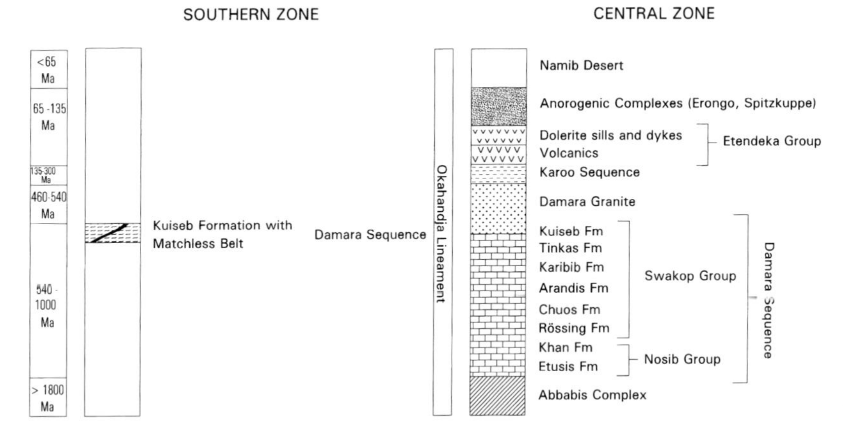


**Fig. 9.1.2:** Stratigraphic column for route 9.1.

ic events. The lineament trends NE and represents a deep penetrating zone of weakness in the crust which has repeatedly been active and has had a major influence throughout the depositional and tectonic history of the Damara Orogen (Miller 1979). The Central Zone is underlain by the low-pressure, high-temperature counterpart of the kyanite belt of the Southern Zone. Arenitic and calc-silicate rocks prevail in the basal Nosib Group which formed in graben structures during intra-continental rifting. Diverse rocks of the overlying Swakop Group were deposited in a spreading ocean on a continental shelf and are subdivided into the Rössing, Chuos, Arandis, Karibib, Tinkas and Kuiseb Formations (see 3.).

On the Okahandja western bypass is the turn-off to the hot springs at Gross Barmen. Here, twelve individual springs occur in quartz-biotite-schist of the Kuiseb Formation at the intersection of a shear zone, which parallels the strike of the schist, and a N-NE trending fault. These tectonic features are related to the same event that produced the Windhoek Graben. The spring water has a very high temperature of 42 °C to 69 °C, and the yield is about 10 m$^3$ per hour. The springs are of therapeutic value because of the dissolved mineral content of the water (Gevers, Hart & Martin 1963).

After Okahandja, the route now follows road B2 in a westerly direction towards Swakopmund, and the morphology of the landscape undergoes a marked change. Some of these changes are associated with a large number of syn- to post-tectonic granite plutons which have intruded the Central Zone during the Damaran Orogeny. Plutonism lasted for 190 Ma starting 650 Ma ago, and the youngest granites were emplaced while the Damara Orogen was undergoing rapid uplift (Miller 1990). The basement of the Damara Sequence in the area is composed of highly metamorphosed rocks of the Mesoproterozoic Abbabis Complex, which are exposed in domes and anti-forms. Magmatic rocks of the Karoo Sequence intrude and Cenozoic sediments of the Namib Desert overlie the Damaran rocks in the W.

About 11 km outside Okahandja the first granites are encountered, and erosion of these plutonic rocks has produced smoothly rounded hills displaying the typical onion-skin weathering. The granite, crosscut by pegmatite dykes, is exposed in a road cut some 15 km outside Okahandja. Thereafter a number of granite hills can be seen in the distance on the northern side of the road. The road crosses numerous ephemeral river beds, which flow towards the Swakop River in the S. One of these, the Ozombanda River, is crossed some 34 km from Okahandja, and to the S this river has incised deeply into Damaran granite, which provides for excellent exposure. Two granites are present: a foliated, syn-tectonic, gray, coarse-grained, porphyritic Salem type granite, which is intruded by dykes of a medium- to coarse-grained reddish granite. The latter has in places developed a pegmatitic texture. As it lacks a foliation, it is therefore interpreted as post-tectonic.

Fifty kilometers outside Okahandja the road reaches Wilhelmstal where a road cut with diorite intruding granite can be observed. The prominent mountain seen to the southern of the road is made up of quartzites of the Nosib Group. As the road continues in a westerly direction, the landscape changes yet again, as the geology consists of marbles of the Karibib Formation of the Swakop Group. The first marble hills can be seen on the southern side of the road some 22 km outside Wilhelmstal and shortly thereafter the mighty Erongo Complex (see 8.7) appears N of the road in the distance. It dominates the landscape to the N until well after Karibib. To the S the undulating marble hills continue. About 50 km from Wilhelmstal two mountains ("Sargdeckel" and "Jungfrau") with quite a different morphology to that of the marble hills, become visible to the S. They consist of Triassic arkose of the Karoo Sequence, topped by hard Cretaceous basalt, which preserved the arkose from weathering.

Three kilometers further on, a marble quarry becomes visible on the southern side of the road on farm Okatjimukuju, and the marble ridge in which the quarry is located continues all the way to Karibib. The Karibib area is the heartland of Namibia's dimension stone industry, and marbles have been quarried here since the beginning of the 20$^{th}$ century. Today's marble production amounts to some 12 000 tons annually, complimented by over 6 000 t of granite. On the eastern fringes of Karibib, the "Berger Marble Crush" company produces various types of fine aggregates from locally quarried marbles. On the western side of the town the road passes the "Karibib Marbleworks" company, where tiles and other dimension stone products are manufactured.

Just 4 km outside Karibib the prominent mountain on the southern side of the road is formed by quartzite and meta-arkose of the Etusis Formation of the Nosib Group, but thereafter, marbles of the Karibib Formation continue to dominate. Namibia's only gold mine, the Navachab Gold Mine, is located in this area and can be seen to the S of the road about 8 km from Karibib. This open pit mine started operations in 1989 and produces about 2 t of gold annually from quartz veins in marble. About 11 km outside Karibib, another marble quarry, the Dernburg Quarry, is visible on the southern side of the road, while the northern view is still dominated by the Erongo Mountain. One km further on, the slimes dam of the Navachab Gold Mine is visible in the mountains to the S, and gives a good indication of the amount of rock that has been mined to recover the gold.

Some 18 km from Karibib the valley narrows, and quartzite and meta-arkose of the Etusis Formation with a core of Abbabis basement form a prominent ridge on the northern side of the road. Isolated outcrops of mica schist of the overlying Arandis Formation, Swakop Group, can be seen close to the road some 22 km outside Karibib, before the descent into the Namib Desert peneplain commences about 24 km from Karibib. Here a first glimpse of the desert with the Spitzkuppe Mountains (see 8.29) can be made through a gap in the mountains to the N of the road. In this region, the road follows the major drainage of the Khan River, a large tributary of the Swakop River, and for this reason the Great Escarpment is not as pronounced as elsewhere in Namibia, mainly due to erosion caused by this drainage, and the fact that the Damaran rocks present here weather more readily than the rocks forming the Great Escarpment elsewhere. S of the road, remnants of calcrete capped terraces can be observed. These were originally deposited by this drainage and then subsequently eroded, as uplift and development of the Great Escarpment continued.

Before entering Usakos, the Nubebberge seen to the S of the road are composed of marble of the Karibib Formation and schist of the Kuiseb Formation. They are extensively intruded by Damaran pegmatites that have been mined for tin and tourmaline since the early days of mining in Namibia. The Usakos Tourmaline Mine is still a major producer of gem quality tourmaline. On leaving Usakos the road crosses the Khan River. It has a huge catchment area in the southern parts of the Erongo Mountain, and large amounts of water are carried by the ephemeral stream during rainy seasons. The river bed is incised deeply, and after crossing the bridge the road rises considerably before reaching the Namib peneplain. The Gamgamichabberge and Otjipateraberge in the distance to the S are made up of quartzite and meta-arkose of the Etusis Formation.

Some 3 km outside Usakos there are road cuts in mica schist of the Swakop Group, and from here, looking back to the E, there is a very good view of the western slopes of the Erongo Mountain. Further on, about 14 km outside Usakos, the Spitzkuppe (see 8.29) appears on the northern side of the road. The fairly flat landscape is now underlain by Damaran granite, however, after about another 3 km the smaller hills along the road mark the transition back into meta-sediments of the Swakop and Nosib Groups. To the S the view across the Khan River Valley shows the Khanberge mountain range composed of metasediments of the Nosib Group which includes a dark amphibolitic layer. Further to the S, the prominent Chuosberge mountain range is made up of Nosib Group metasediments as well as Chuos Formation schist and Karibib Formation marble.

As the road passes the distant Spitzkuppe, even the Brandberg (see 8.1) can be seen in the NW on a clear day. Here, the first dolerite dykes can be observed on the southern side of the road. The occurrence of dolerite dykes increases towards the coast and they become a dominating feature of the landscape closer to Swakopmund. Dolerite dykes form NE, and less frequently E-W trending dyke swarms, which cut through all the meta-sediments and intrusive rocks. The dyke swarms are related to the Cretaceous Etendeka volcanism, which immediately preceded the break-up of the African-South American part of Gondwanaland. Individual dykes range between 10 cm and several tens of meters wide, and often extend for many kilometers along strike. They form prominent ridges because of their relative hardness compared to the country rocks, which they intruded. The dark-brown, fine-grained dolerite consists mainly of large plagioclase and olivine phenocrysts in a groundmass of fine-grained plagioclase and opaque minerals.

To the S of the road, some 46 km outside Usakos, there is another particularly good view of the Khan River Valley, where layering in Swakop Group meta-sediments is exposed. Also here, on the northern side of the road dolerite dykes can be seen which have accumulated wind-blown sand of the Namib Desert on their southwestern sides due to the prevailing southwesterly winds.

Sixty kilometers from Usakos and from there onwards, it can be observed in the Khan River Valley to the S, that the Swakop Group meta-sediments become increasingly intruded by granite, alaskite and pegmatites. Some of the alaskites carry uranium and one such alaskite is mined at Rössing, Namibia's only uranium mine. The Rössing Mine, which is situated close to the road 90 km from Usakos and 60 km from Swakopmund, is the largest open pit uranium mine in the world and has the capacity to produce some 5 000 short tons of uranium oxide per year. Regular visits to the mine are organised by the Swakopmund Museum. Between Rössing and Swakopmund a water pipeline carrying the water supply of the Rössing Mine follows the road on its southern side, and it is noteworthy that the water for this mine, situated in a desert environment, needs to be transported for over 100 km from its source in the Kuiseb River close to Walvis Bay.

Shortly after Rössing Mine, there is a turn-off marked "Khan Mine Gorge". The Khan Copper Mine exploited a rare copper-bearing pegmatite in the early days of the 20$^{th}$ century, and was once one of the largest mines in the country. Some 100 km from Usakos the prominent Rössing Mountain appears in the N. This inselberg is composed of calc-silicates of the Khan Formation, Nosib Group, which have been quarried for road and railway ballast. Just past the Rössing Mountain a road turning off to the N leads to the now closed Namib Lead Mine, an underground lead-zinc mine which produced some 26 000 t of lead and 27 000 t of zinc between 1969 and 1991. The ore occurs in the form of massive sulphide bodies as well as a network of stringers in marble and schist of the Karibib Formation.

The road now follows the Swakop River Valley until it reaches Swakopmund. The landscape in the valley is dominated by abundant dark dolerite dyke ridges, which are in a stark contrast to the light-coloured sand. Because of this, as well as the rugged topography of the deeply incised Swakop River Valley and its tributaries and the almost complete lack of vegetation the area is also called "Moon Landscape".

Some 20 km before Swakopmund is reached, the Swakop River Valley appears on the southern side of the road. Irrigation schemes have been established in the valley, and water is pumped from the sands of this mighty ephemeral river. The southern bank of the river is dominated by dolerite dykes and sand, moving in from the S. About 2 km further on, a dark dolerite plug forms a very prominent hill S of the river. At Nonidas, 15 km outside Swakopmund and N of the road, a quarry has been established in dolomitic marbles of the Chuos Formation. The marble contains green serpentinite and brown siderite, and this colourful appearance gives it the trade name "Namib Harlequin".

Just outside Swakopmund, the coastal dune belt appears to the S. This coastal dune belt occupies an approximately 5 km wide coastal strip between Walvis Bay and Swakopmund. The high sand dunes represent an extension of the Namib Sandsea which is otherwise confined to the area S of the Kuiseb River. Regular floods of the Kuiseb River prevent the dunes from extending across its course, however, since the Kuiseb River only reaches the sea in those years with exceptional rainy seasons, the prevailing southwesterly winds are able to blow sand northeastwards in the coastal area.

## 9.2 Karibib – Omaruru – Uis

by Volker Petzel

The route from Karibib to Omaruru and further to Uis is 165 km long, and follows a half-circle around the prominent Erongo Mountain. It starts off in rocks of the Central Zone of the Damara Orogen, and later crosses the lineament forming the northern Boundary of the Zone, to end off in rocks of the Northern Zone of the Damara Orogen (9.2.1, 9.2.2) The C33 turn-off to Omaruru is located some 3 km to the E of Karibib on the main road B2. The first prominent range of hills, called Affenberg, occurs some 4 km from the turnoff on the eastern side of the road. The rocks outcropping along this ridge belong to the Nosib Group and comprise feldspathic quartzite of the Etusis Formation and thinly bedded schist and calc-silicate of the overlying Khan Formation. Some 7 km further to the N, the road transects an outcrop of marble of the Karibib Formation, Swakop Group, which forms a prominent ridge to the E on farm Okawayo. Another prominent ridge is cut by the road some 21 km from Karibib. Before it descends into the Khan River and mica schist, impure marble and calc-silicate bands of the Karibib Formation are exposed in road cuts. Along the descent into the Khan River, a N-NE trending pegmatite dyke, the Etiro Pegmatite, forms a prominent ridge within a dome of coarsely porphyritic biotite granite. The Etiro Pegmatite was mined in the past, predominantly for beryl.

The various units forming the Erongo Mountain (see 8.7) can also be seen from the Etiro Pegmatite. At the base smoothly weathered Damaran granite, intrusive into Kuiseb Formation schist, is exposed. Karoo sediments of the Lion's Head Formation unconformably overlie the granite and the schist. The Karoo sediments have a distinct horizontal layering and consist of poorly sorted conglomerate, quartzite and arkose. The distinct vertical cliff face of the Etiromund Member can be seen near the top. This

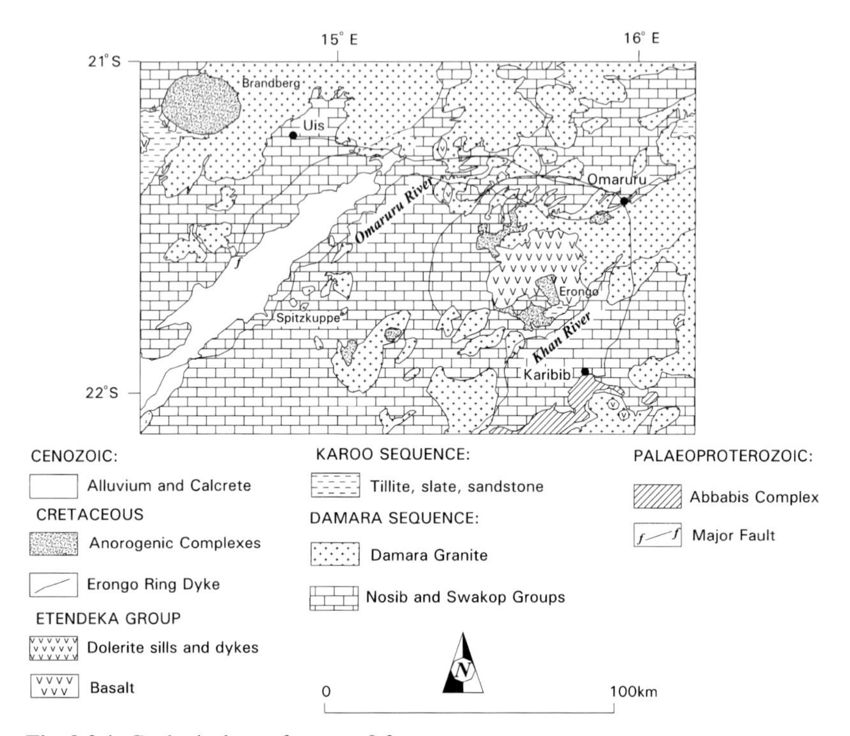


**Fig. 9.2.1:** Geological map for route 9.2.

unit consists of gritstone and forms the boundary between the lower and middle units of the Lion's Head Formation. The sediments are unconformably overlain by basaltic lava flows of the basal Etendeka Group, and exposed at the top of the mountain are rhyolitic ash-flow tuffs of the Ombu Member.

The Giftkuppe, a prominent granite inselberg, can be seen to the W of the road some 32 km from Karibib. It consists of albitised granite, which has been intruded by sheeted pegmatite dykes, and during 1936 rutile was mined from greisen zones located in the granite. Several outcrops of syn- to post-tectonic Damaran granites are traversed on the remainder of the road to Omaruru. A prominent granite inselberg, the Omaruru Kuppe, can be seen SE of the town, and is the symbol of the town. The Omaruru River, one of the largest ephemeral rivers in Namibia, is crossed on the southern fringes of the town.

In the center of Omaruru, road C36 to Uis turns off to the NW. Following this road, the Erongo Complex lies to the S. Among the foothills on the northeastern side of the com-

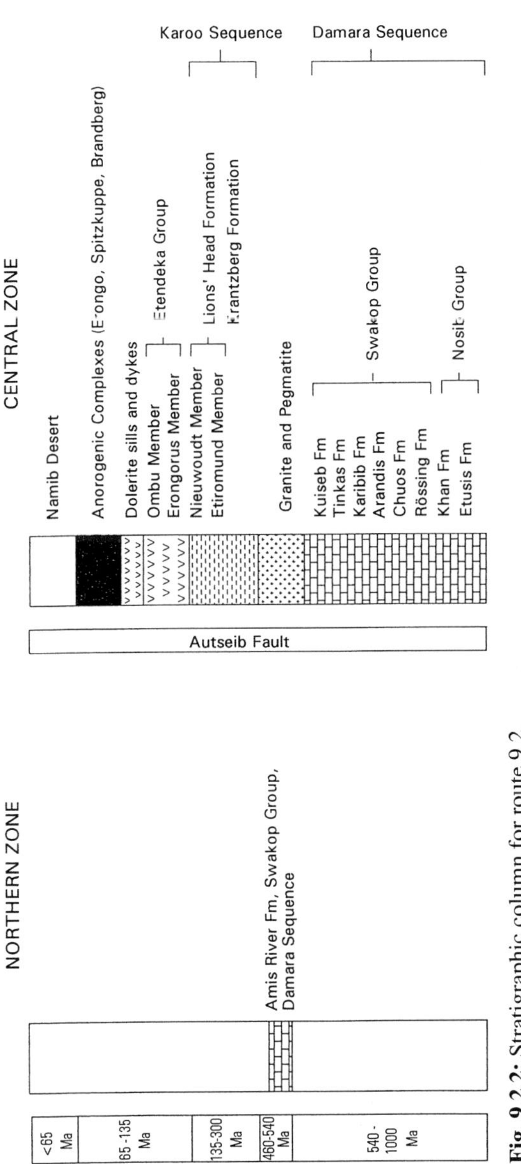


**Fig. 9.2.2:** Stratigraphic column for route 9.2.

plex, the Krantzberg forms a prominent morphological feature. The base of this hill consists of Kuiseb Formation quartz biotite schist intruded by Damaran granite, which is capped by Karoo conglomerates and arkoses of the Krantzberg Formation. The sediments are in turn overlain by basaltic lavas of the Erongo Complex. Some of the mine workings of the now defunct Krantzberg Mine can be seen just below the Krantzberg Formation. This mine was the major producer of tungsten in Namibia during the 1940s and into the mid 1950s. Greisen zones developed during the intrusion of the Erongo granite contain wolframite, fluorite, beryl, and associated minor sulphide mineralisation. The greisen bodies occur predominantly near the contact of the Damaran granite and the schist, but also in places extend into the overlying Karoo lithologies.

Opposite the Krantzberg Mine, a range of hills, the Elephantenberg, looms on the horizon towards the N. This mountain predominantly consists of marble of the Karibib Formation.

An olivine-dolerite ring dyke, related to the Erongo Complex, occurs within a circular fracture to the N and W of the mountain, some 20 km from the volcanic center (see 8.7). This dyke is traversed several times by the road and can best be inspected close to a river some 21 km from Omaruru, where rounded boulders of the dyke are present next to the road. The olivine dolerite has a modal composition of 67 % labradorite, 25 % augite, 5,7 % olivine and 2 % opaques and other minor minerals. The magnetic signature of this dyke indicates that it is inclined towards the complex, and might therefore be a cone sheet rather than a ring dyke (Pirajno 1990).

Fourtysix kilometers from Omaruru, the Okombahe Mountain, which predominantly consists of olivine dolerite of the Karoo Sequence, appears in the Omaruru River depression to the SW. Further to the SW, the post-Karoo granites of the Grosse and Kleine Spitzkuppe form prominent inselbergs in the distance. Just past the Okombahe turn-off, the Paukuab range occurs on the northeastern side of the road. These hills chiefly consist of an olivine dolerite dyke which intruded Damaran granites. To the W of the road towards Uis, a white range of hills consisting of marbles of the Karibib Formation can be seen in the Omaruru River depression. The small, roundish rock outcrops in the foreground consist of Damaran granite.

Closer to Uis, the Autseib Fault, marking the transition from the Central Zone to the Northern Zone of the Damara Orogen, is crossed, however, this can hardly be noticed. Meta-greywackes of the Amis River Formation, Swakop Group of the Damara Sequence are exposed at Uis. Pegmatite swarms containing cassiterite and tantalite mineralisation within greisen zones cut the meta-greywackes and were mined at Uis from 1911 until 1988 when the mine had to close due to the collapse of the tin price and the high operational costs related to the low grade of the ore. When in operation, the Uis Tin Mine was the world's largest hard rock tin mine; it still has ore reserves of 72 million t grading 0,1 – 0,15 % Sn.

# 9.3 Windhoek – Bosua Pass – Swakopmund

For almost half of its length, including the descent down the Great Escarpment, this 332 km long route transects the thick sequence of Kuiseb Formation schist of the Damara Sequence, then passes different Damaran granites, crosses the coastal plain of the Namib Desert and finally reaches the Atlantic coast at Swakopmund (Figs 9.3.1, 9.3.2).

The route leaves Windhoek in a westerly direction on road C28, which for the first 25 km climbs out of the Windhoek valley, in Kuiseb Formation schists, which are well exposed in a number of road-cuts. The Kuiseb Formation is an almost 10 000 m thick sequence of intercalated mica-schist, quartzite and meta-greywacke which has been deposited in a spreading ocean, and has undergone metamorphism during the Damaran Orogeny. Faulting has facilitated the incision of deep valleys that contribute water to the Swakop River drainage in the N, and one such valley, the valley of the Aretaragas River, is crossed after 12 km.

The amphibolites of the Matchless Belt can be seen in road-cuttings some 8 km and 9 km outside Windhoek, and the belt is once again crossed by the road 14 km outside Windhoek. The Matchless Amphibolite Belt is a linear feature up to 3 km wide and some 350 km long. It consists of amphibolites, which form lenses and layers interbedded with the Kuiseb schist. The amphibolites have MORB geochemical signature, and have therefore been interpreted as metamorphosed syn-sedimentary submarine volcanics emplaced into sediments covering a mid-ocean ridge that had developed in the Damaran Southern Zone ocean. The amphibolites are associated with important mas-

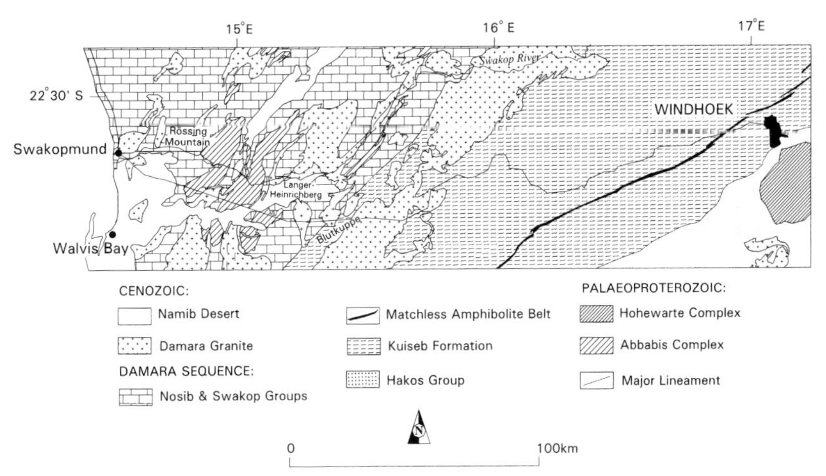


**Fig. 9.3.1:** Geological map for route 9.3.

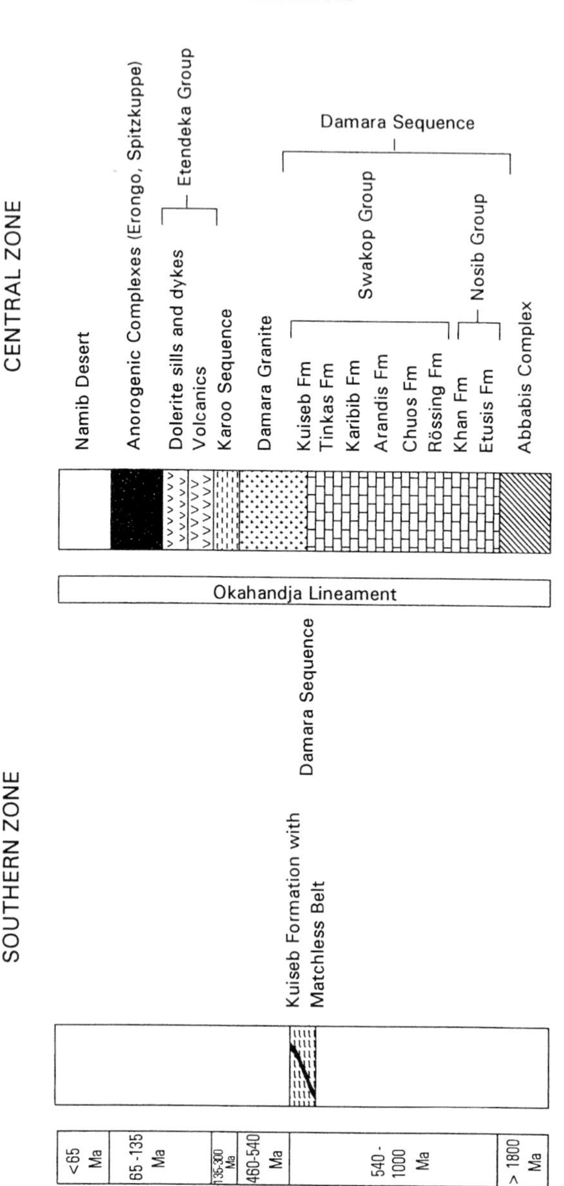


**Fig. 9.3.2:** Stratigraphic column for route 9.3.

sive sulphide deposits such as the Matchless Mine, the turn-off to which is reached after another 8 km, just after the tarred road ends. The now abandoned Matchless Mine is one of the oldest mines in Namibia and started production as early as 1852. The old workings are of historic interest and are a proclaimed National Monument. The mine was active intermittently until 1983, during which time more than 30 000 t of copper were produced.

After the turn-off to the Matchless Mine, the road has effectively reached the plateau of the Khomas Highland and follows an undulating peneplain with almost no outcrops, apart from cuts in riverbeds, which the road crosses from time to time. The Kuiseb schist dips at 45° to the NW and can be observed in these rare outcrops. Some 76 km from Windhoek, the prominent flat-topped Gamsberg (see 8.10), situated on the edge of the Great Escarpment, can be seen to the SW.

After another 30 km the view to the W opens up, and, apart from the surrounding undulating Kuiseb schist hills, the Donkerhuk Granites W of the Great Escarpment become visible. The Donkerhuk Granite is a grey, garnet-bearing, two mica, post-tectonic Damaran granite, which intruded the Kuiseb schist some 520 Ma ago. Just 3 km further on, the top of the Bosua Pass is reached, which provides a magnificent view across the Great Escarpment into the lowlands to the west. The very steep descent down Bosua Pass is only 2 km long, with a gradient of 1:5. Kuiseb schist is well exposed in roadcuttings throughout these 2 km. Having reached the bottom of Bosua Pass, the landscape is still dominated by the schist although the prominent massif of the Witwatersberge, composed of Donkerhuk Granite, appears to the N of the road.

Ten kilometers from the bottom of Bosua Pass, more Donkerhuk granite is visible to the N of the road, while the road itself is still leading through terrane underlain by Kuiseb schist, and the mountain range to the SE is also composed of Kuiseb schist. The schist dips steeply to the NW. As the road continues, the valley widens, and after 17 km Donkerhuk Granite exposures appear closer to the road in the north. On farm Donkerhuk East, a prominent granite hill on the northern side of the road shows the intrusive contact with the Kuiseb schist at the base. Three kilometers further on, granite also appears to the S. As the valley narrows, the road enters into terrane underlain by granite at the western boundary of farm Donkerhuk West. For the next 10 km the road winds itself through a narrow valley flanked by granite hills displaying typical onionskin weathering. The coarse-grained texture of the granite can be observed in outcrops right next to the road. Just after the road 1985 turn-off to the S, the intrusive relationship between the granite and the schist is well exposed to the N of the road.

Further on, the valley opens up again, but the landscape is still dominated by the typical morphology of the granite mountains flanking it on both sides. After 5 km the road passes a dry riverbed, the course of which is marked with large trees. Amongst them are many quiver trees, and it is remarkable how these have utilised the exfoliation cracks in the granite to sink their roots and utilize the moisture available in the cracks. The road enters the Namib Naukluft Park after another 6 km, leaving behind outcrops of Damaran rocks and leading onto a vast peneplain where Cenozoic sediments of the Namib

Desert form a thin cover to the Proterozoic rocks. The peneplain is dissected by numerous dry riverbeds, which dewater into the Swakop River. The gradual, gentle gradient of the landscape towards the Swakop River Valley can be well appreciated to the NW of the road for the next 20 km. Calcrete capping the Namib sediments can be observed in incisions made by the small rivers draining into the Swakop River valley. Calcrete is a hard surface limestone common in warm, arid and semi-arid regions where it builts up by solution and re-deposition of lime by meteoric waters. The lime commonly cements sand and gravel beds, and can even partly replace older sediments.

As the road continues, one of the most important boundaries in the Damara Orogen, the Okahandja Lineament is crossed. This lineament represents the southern edge of the Central Zone of the Orogen, and marks major changes in stratigraphic succession, structural style and age of metamorphic events. The lineament trends N and represents a deep penetrating zone of weakness in the crust which has repeatedly been active and which had a major influence throughout the depositional and tectonic history of the Damara Orogen (Miller 1979, 1983). However, there is no marked change in the landscape.

The turn-off to Blutkuppe to the N is reached 24 km after entering the Namib Naukluft Park, in an otherwise very plain landscape, which is only interrupted by lines of small trees marking the courses of dry riverbeds. Blutkuppe is a prominent mountain composed of gray to pink, porphyritic biotite granite belonging to the post-tectonic suite of Damaran granites with an age of approximately 520 Ma. The even more prominent dark massif to the N of Blutkuppe is Langer Heinrichberg. This massif contains quartzite and meta-arkoses of the Etusis Formation of the Nosib Group at the base of the Damara Sequence. The meta-sediments of the Etusis Formation also form the dark mountain range produced by an anticlinal structure to the N of the road some 7 km after the Blutkuppe turn-off.

From there, for the next 7 km the landscape is underlain by a Damaran granodiorite, and rounded boulders of that material produced by weathering can be seen strewn over the peneplain. The mountain range to the S of the road is composed of schist of the Tinkas Formation, Swakop Group, which has been intruded by the granodiorite some 520 Ma ago.

As the road passes over an elevated area, outcrops of the Tinkas Formation schist can be observed right next to the road for the next 5 km. The road then leads onto the Namib peneplain again, and after another 7 km gneisses of the Palaeoproterozoic Ababis Complex form smooth little hills on either side of the road for some 6 km. Thereafter dolerite ridges make their appearance in the desert plain. They mostly strike in a coast-parallel direction and towards the coast they increase in abundance. The dyke swarms are related to the Etendeka volcanism and the break-up of Gondwanaland. Individual dykes are between 10 cm and several tens of meters wide, and often extend for many kilometers along strike. Because of their relative hardness compared to the country rocks into which they intruded, they form prominent ridges. The dark-brown, fine-grained dolerite consists mainly of large plagioclase and olivine phenocrysts in a groundmass of fine-grained plagioclase and opaque minerals.

To the N of the road the relief is such that one can see as far as the Erongo Mountains (see 8.7) in the N. Smaller hills to the S of the road are composed of white marbles of the Karibib Formation. For the next 30 km the landscape is dominated by a desert plain dotted with small hills composed of granites as well as calc-silicate and schist of the Nosib Group, both of which are intensely intruded by dolerite dykes. The road then passes a range of marbles of the Arandis and Karibib Formations, before the landscape looses relief and there are no further outcrops, although run-off patterns towards the Swakop River drainage are very apparent.

After some 12 km there is a turn-off to the N to the so-called "Moon Landscape". The Moon Landscape is a rugged terrane formed by the deep incision of the Swakop River into meta-sediments of the Chuos, Rössing and Khan Formations of the Damara Sequence. These meta-sediments are intensively intruded by dolerite dykes, which, in the prevailing arid climate, are more resistant to weathering and have therefore formed prominent, dark ridges. The area, which has extremely little vegetation, together with the dramatic scenery of the deeply incised valley has inspired the name "Moon Landscape".

The western boundary of the Namib Naukluft Park is passed just after the turn-off to the Moon Landscape, and after another 3 km one can see the coastal dune belt. This coastal dune belt runs in a coast-parallel direction from Walvis Bay to just S of the Swakop River. It is the only place where the Namib Sandsea extends northwards across the Kuiseb River. While floods of the Kuiseb flush out the windblown sand further to the E and thereby prevent extensions across its course, the fact that the river usually does not reach the ocean has allowed for the establishment of this narrow coastal dune belt. Some of the highest sand dunes in Namibia, such as Dune 7, are located here. Noteworthy are the black areas on the dune surfaces, which represent wind-generated concentrations of dark heavy minerals such as magnetite, ilmenite and garnet.

Six kilometers after leaving the Namib Naukluft Park, one can see Rössing Mountain prominently to the N. This major anticlinal structure in more than 1000 m thick, bedded, grayish-green calc-silicate rocks of the Khan Formation, Nosib Group, forms a prominent inselberg in the Namib Desert. Thereafter the decent into the valley of the Swakop River starts. The river has incised into a number of terraces, which are passed before the actual river is crossed 5 km before reaching Swakopmund. Abundant reed beds indicate the permanent moisture available in the sands of the Swakop River. To the W, the railway bridge is built on white, coarse-grained and graphite-bearing marble of the Karibib Formation. The road then climbs out of the Swakop River valley into the coastal plain where it reaches Swakopmund.

# 9.4 Windhoek – Gamsberg-Pass – Walvis Bay – Swakopmund

This 395 km long route starts in the highlands S of Windhoek, follows the belt of thrusting along the southern margin of the Damara Orogen, descends down the Great Escarpment and passes the Kuiseb River valley, it then crosses the Namib Desert to reach the coast at Walvis Bay (Figs 9.4.1, 9.4.2). Leaving Windhoek southwards on road C26 to the S, the road winds out of the Windhoek valley via the Kupferberg Pass onto the upland plateau through the Kuiseb Formation of the Damara Sequence. The Kuiseb Formation is an almost 10 000 m thick sequence of intercalated mica-schist, quartzite and meta-greywacke which has been deposited in a spreading ocean, and has undergone metamorphism during the Damaran Orogeny. The prominent Auas Mountains can be seen to the E. These mountains were formed by intense thrusting during the same orogeny, and consist chiefly of Auas Formation quartzite of the Hakos Group, Damara Sequence.

For the first 13 km there are numerous road-cuts in which schist and a little quartzite can be observed, as well as large quartz segregations that occur within the schist. After

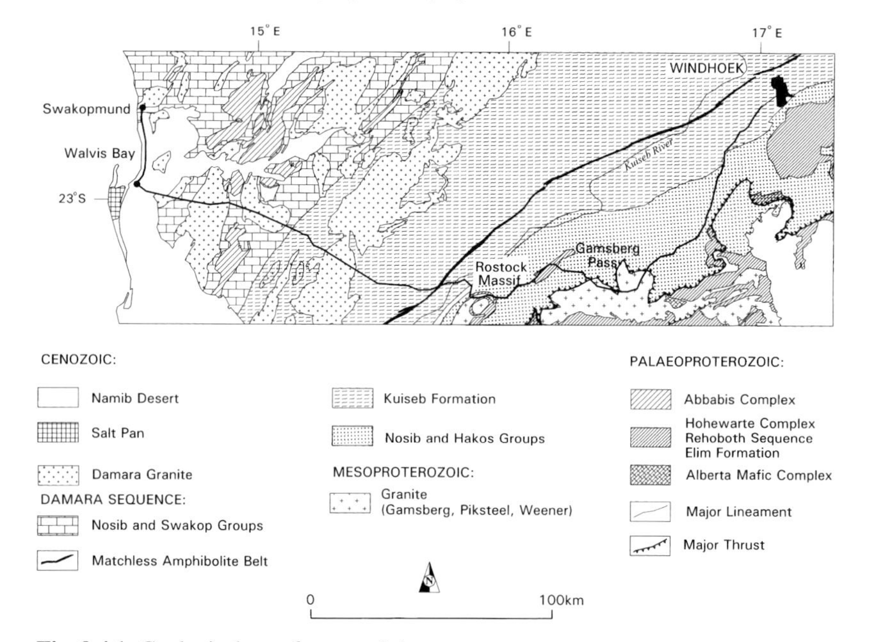


**Fig. 9.4.1:** Geological map for route 9.4.

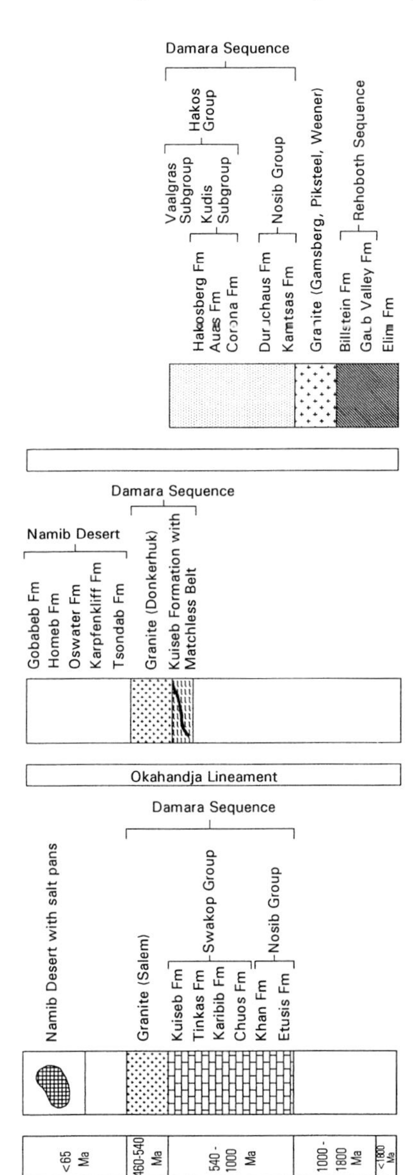


**Fig. 9.4.2:** Stratigraphic column for route 9.4.

descending into the deeply incised valley of the Aretaragas River, the road starts climbing the Kupferberg Pass about 8 km outside Windhoek. The top of the pass is reached 18 km from Windhoek, and from here, one has a magnificent view of the highlands to the S of Windhoek (Fig. 9.4.3). The peneplain to the SW is underlain by Kuiseb schist, and is dotted in the E with a number of inselbergs composed of erosion-resistant quartzite of the Auas Formation. At the top of Kupferberg Pass, the road passes a major water divide, with rivers in the Windhoek valley dewatering into the Atlantic drainage via the Swakop River, while the watercourses on the upland plateau drain into the Kalahari Basin to the E, and some of them eventually into the Orange River.

Two kilometers further, the prominent flat-topped Gamsberg (see 8.10), situated on the edge of the Great Escarpment, can be seen in the SW. The road thereafter follows the peneplain, and although there is very little outcrop, the surface is strewn with rubble derived from the quartz segregations in the schist. At a junction, some 30 km outside Windhoek, road C26 turns to the S, and the landscape is now underlain by schist and minor quartzite of the Vaalgras Subgroup. For the next 160 km the geology is dominated by rocks of the Nosib and Hakos Groups of the Damara Sequence, which were deposited on what was the northern edge of the Kalahari Craton. An earlier rift (Nosib Group) was followed by the development of an Atlantic-type passive continental margin (Hakos Group). The passive margin rocks underwent extremely complex deformation during the subsequent mountain-building process.

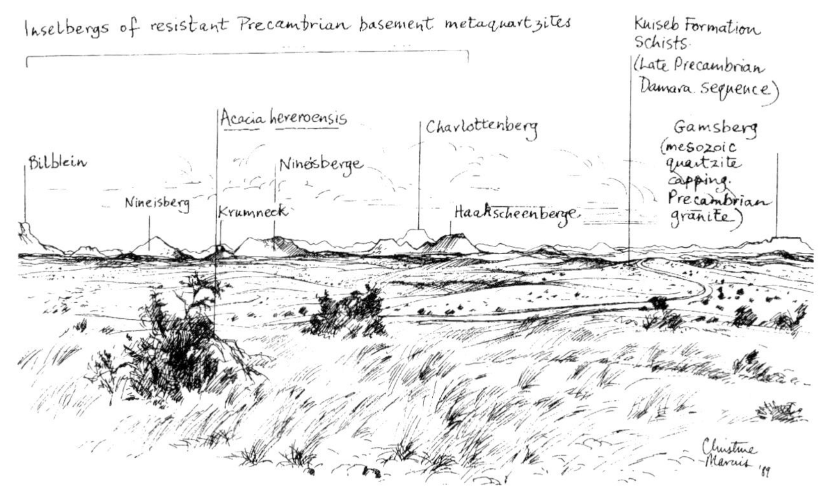


**Fig. 9.4.3:** View from top of the Kupferberg Pass (by Christine Marais).

A quartzite in the Vaalgras Subgroup can be clearly seen in a road cut some 2 km from the road junction and the hills NW of the road are also composed of this quartzite. Schist and diamictites of the Chuos Formation, deposited under glacio-marine conditions, soon thereafter underlie the peneplain. However, there is hardly any outcrop. Some 16 km further on the road passes through a slightly more hilly landscape, and Charlottenberg, where Palaeoproterozoic quartzite of the Billstein Formation has been thrust over Damaran meta-sediments, appears in the S. The Haris River winds itself through the peneplain to the N of the road, before it links up with the Gurumanas River.

Two kilometers further on, a prominent mountain range in the SE is geologically similar to Charlottenberg. The road then follows a straight southerly course in an undulating landscape, which is underlain by schist and quartzite of the Duruchaus Formation of the Nosib Group associated with a grassy vegetation otherwise not present. The Duruchaus Formation formed during the early rifting stages under saline conditions in a playa-lake environment, similar to today's East African Rift Valley. After 14 km, the road passes the Gurumanas River, and the quartzitic schist of the Duruchaus Formation can be seen in rare outcrops on both sides of the road (Fig. 9.4.4). Some hills composed of schist and marble of the Kudis Subgroup, Hakos Group display large quartz segregations to the W of the road.

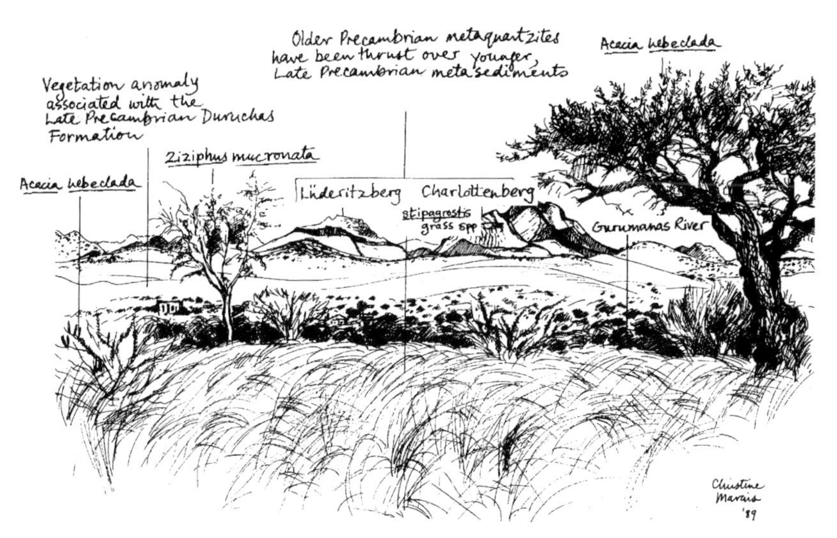


**Fig. 9.4.4:** View over the Gurumanas River (by Christine Marais).

The road then crosses a very flat peneplain underlain by meta-sediments of the Chuos Formation, and for the next 20 km only dry riverbeds, which dewater towards the SE, provide any topographical relief. Thereafter the topography increases once more with quite a number of prominent dry river beds and schist of the Chuos Formation is seen in road cuts and form the small hills in the area. Soon after, the landscape becomes flat, and dominating the view are the Mesoproterozoic granitic mountains of the Great Escarpment to the SW, with the Gamsberg very prominent in the W.

After another 21 km, the road takes a marked turn to the W, in an area that is underlain by young sediments on the farm Göllschau. The prominent Kamelberg mountains to the S are composed of amphibolites of the Elim Formation and along the Kamelberg thrust to the SE, the Elim and Gaub Valley Formations have been thrust over Gamsberg granite. Just 2 km further on, the road enters terrane underlain by Gamsberg granite, which can be observed in numerous small outcrops. The texture is typical of granite, and weathering produces the red sand visible on the road and after 3 km, still on farm Göllschau, well developed onionskin weathering of the granite can be observed. Thereafter the road moves into dolomites of the Corona Formation on the farm Weissenfels, and further on, dark schist with intercalated banded iron formation of the Chuos Formation are encountered. They dip at an angle of 45° to the NW and weathering has produced conspicuous outcrops with harder material protruding at that angle from the surface.

Some 10 km further on, the road passes back into Gamsberg Granite terrane, and small hills to the N of the road are also composed of this material. However, after 5 km a road-cutting to the S of the road on the farm Hakos shows fine-grained reddish meta-sediments of the Kudis Subgroup, which now form the country rock, followed shortly thereafter by rocks of the Chuos Formation. After the farm boundary, one gets a first view of the Hakos Mountains to the NW. Just one more kilometer, and the top of the Gamsberg Pass is reached with its magnificent view of the Hakos Mountains and in the distance the yellowish sand sea of the Namib Desert (Fig. 9.4.5). The dissected Great Escarpment in this area represents the deeply eroded roots of the Hakos Mountains thrust sheets. The Hakos Mountains are composed of turbiditic quartzite of the Kudis Subgroup. Their spectacular morphology is related to their position within the Southern Zone thrust belt, and the associated intense thrusting that has taken place in the area during the formation of the Damara mountain belt.

For the next 8 km the Gamsberg Pass descends steeply down the Great Escarpment, and there is continuous outcrop of Chuos Formation schist along the road. After 3 km a mylonitic zone and a large quartz vein in the schist can be observed. Once the bottom of the Gamsberg Pass is reached, the road follows the valley of the Djab River in the Hakos Mountains. The road crosses the dry riverbed frequently, and many good outcrops are seen. After 16 km a major thrust is crossed, and quartzites of the Kudis Subgroup now form the outcrops next to the road. Before the road takes a marked turn to the SW, the bed of the Djab River is once again passed, and is partly underlain by the Mesoproterozoic Piksteel granodiorite and the Weener quartz-diorite. The valley opens

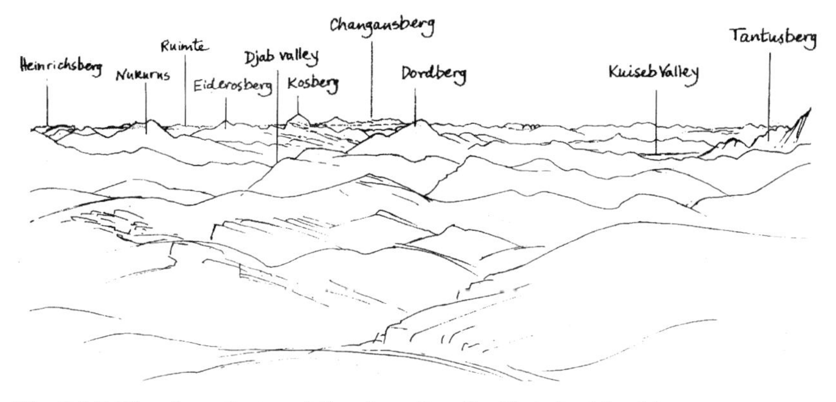


**Fig. 9.4.5:** View from the top of Gamsberg Pass (by Christine Marais).

some 10 km further on, and outcrops next to the road become somewhat rare, although the surrounding mountains are still composed of Kudis Subgroup meta-sediments (Porada & Wittig 1976). After the Djab River crossing, some 20 km from the bottom of Gamsberg Pass, young, red sand dunes can be seen in the now rather wide valley and the name of the farm, "Rooisand", therefore comes as no surprise.

Just after Rooisand, the road passes another major thrust, and the valley is now underlain by turbidites of the Hakosberg Formation. These deep-water sediments are interpreted as having been derived from the quartzites of the same formation, which had formed a sand ridge on a marine fan. A number of thrust planes can be seen to the SE of the road and give a good impression of the intense tectonics that were active in this part of the Southern Margin of the Damara Orogen. From here on, the valley continues to open, and the flanking mountains become lower. Twenty kilometers from Rooisand, in the distance the Namib Desert can be seen to the SW. To the S of the road and continuing for about 7 km, younger deposits of the Sandsteenberge, comprising 10 to 20 Ma old Tsondab Sandstone occur and are capped by calcrete, which gives them their flat-topped appearance (Fig. 9.4.6). Calcrete is a surface limestone common in warm, arid and semi-arid regions where it builts up by solution and re-deposition of lime by meteoric waters. The lime forms a hard cement of sand and gravel beds, and can even partly replace older sediments.

Finally the valley links up with the Namib Desert and young sediments form the underlying rocks. Hakosberg Formation rocks can only be seen in the distance to the N. S of the road, the impressive Rostock Massif, composed of thrust sheets of Palaeoproterozoic schist and amphibolite of the Gaub Valley Formation and the Mesoproterozoic Gamsberg Granite, dominates the landscape. After another 6 km, the road meets the

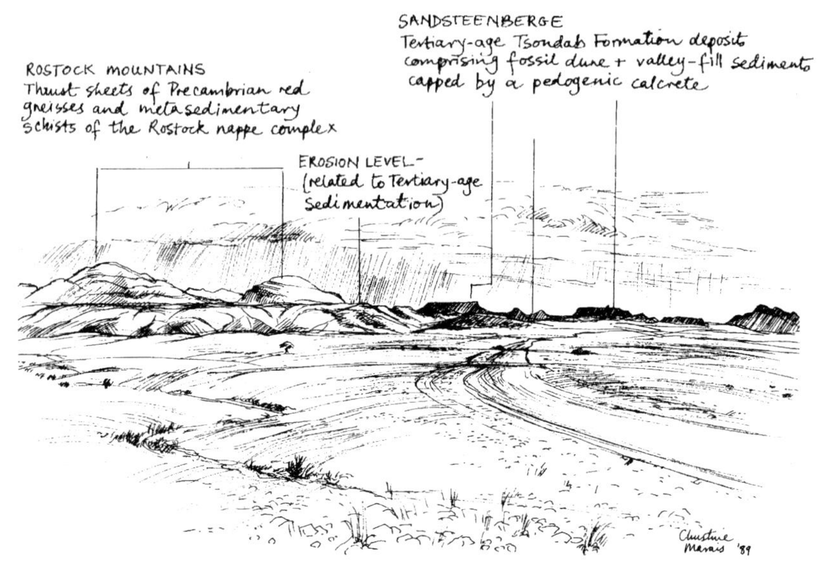


**Fig. 9.4.6:** View over the Rostock Mountains (by Christine Marais).

C14 road from Walvis Bay to Solitaire, and the route continues in the direction of Walvis Bay.

In a road cut, 3 km after the turn-off, schist of the upper Hakos Group is capped by a thick layer of calcrete. In the desert, to the SW, the prominent Kammberg, composed of meta-sediments of the Hakos Group and some lower hills comprising Kuiseb Formation schist can be seen. The Tertiary age Karpfenkliff conglomerate, a proto-Kuiseb River sediment that is usually cemented by calcrete, which is well exposed 6 km after the turn-off, thereafter underlies the area. Outcrops are, however, too small to be shown in Fig. 9.4.1.

To the N the Changansberg mountain, composed of Kuiseb schist, can be seen in the distance. 10 km after the turn-off, just before entering the Namib-Naukluft-Park, the road ascends and cuts into Kuiseb schist. This marks a major divide between the Southern and Southern Margin Zones. The Rostock Massif is from here again prominently visible in the S and further on, tributaries of the Kuiseb River increasingly dissect the landscape. This provides for good outcrops in the schist and the marked dip to the NW is well exposed. After another kilometer, the decent into the Kuiseb Valley begins. To the S, young sediments of the calcrete-cemented Karpfenkliff conglomerate form a conspicuous hill, and a lookout point to the S of the road provides a panoramic view into the Namib Desert. As the road continues, road-cuts provide massive outcrops in schist of the Kuiseb Formation.

The Kuiseb River is reached 3 km after the signpost "Kuiseb Pass", and is crossed by a bridge. At this crossing, the river is in a steep canyon incised into schist of the Kuiseb Formation. The Kuiseb River is one of the major ephemeral watercourses in Namibia dewatering into the Atlantic Ocean (see 8.16). After good rainfalls in the catchment area, flood events occur, however, they rarely reach the sea. The course of the Kuiseb River downstream from the Kuiseb Pass demarcates the northern boundary of the main Namib Sand Sea. The subterranean water reserves support a well-developed flora and associated fauna across the otherwise hyper-arid central Namib Desert.

After the bridge, the road continues for a short while in the valley of the Rutile River, a tributary of the Kuiseb River and spectacular outcrops of Kuiseb schist can be seen just next to the road. The road crosses the Rutile River 2 km further on, and as the road climbs out of the valley, it crosses the amphibolites of the Matchless Belt, which are well exposed in road-cuts on both sides of the road. The Matchless Belt is a linear feature up to 3 km wide and some 350 km long, consisting of amphibolites, which form lenses and layers inter-bedded with the Kuiseb schist. The amphibolites have MORB geochemical signature, and have therefore been interpreted as metamorphosed syn-sedimentary submarine volcanics, emplaced into sediments covering a mid-ocean ridge that had developed in the Damaran southern zone ocean. The amphibolites are associated with important massive sulphide deposits such as the Otjihase and Matchless Mines. They are well exposed on both sides of the road.

Four kilometers after the Rutile River crossing, the road having left the valley, crosses a peneplain underlain by Kuiseb schist, partly covered by Tsondab sandstone. This Tertiary sediment, with its thick calcrete capping, can be observed to the S of the road. There are isolated mountains in the W, which are composed of Kuiseb schist and granite intrusive into the schist. After another 3 km there is a turn-off to the S, which leads to the lookout point into the Kuiseb Canyon (see 8.16). A permit is required for this road, but it is a very worthwhile detour providing a magnificent view into the canyon. The Kuiseb Valley here contains extensive Cenozoic sediments, which have helped to understand the geological evolution of the central Namib region. The stratigraphy of the Kuiseb Valley can be divided into two stages relative to canyon incision. A pre-incision suite includes the Early to Middle Tertiary Tsondab Sandstone and the Karpfenkliff Conglomerate. The post-incision suite comprises the Late Tertiary to Quaternary Oswater Conglomerate, the Homeb Silts and the Gobabeb Gravels. Outcrops of these units are, however, too small to be shown on Fig. 9.4.1. At the lookout point, the canyon can be seen deeply incised into a sequence of Tsondab Sandstone, capped by Karpfenkliff Conglomerate.

A mention must be made here of Henno Martin and Hermann Korn, two famous geologists who took refuge in the Namib for more than 2 years during World War II. The Kuiseb Canyon was one of their hideouts. Their sojourn is described in the book "The sheltering desert" (Martin 1970).

After the lookout, the road continues in areas underlain by Tsondab Sandstone, and the thick calcrete capping can be seen on either side of the road, but after 6 km Kuiseb schist re-appears in outcrops on both sides of the road. The geology from here and for the next 40 km is dominated by hills of Kuiseb schist and a peneplain covered by the Cenozoic sediments of the Namib Desert. It is worth noting, that the hills accumulate wind-blown sands on their eastern slopes from the prevailing winds. Passing the turn-off to Zebra Pan and Ganab, the Namib surface becomes very flat and is characterised by calcretisation. A lookout-point 7 km from the turn-off provides a nice view of Kuiseb schist hills and the desert plain with trees marking otherwise invisible small ephemeral river courses (Fig 9.4.7). Tall Kamelthorn trees mark a larger one of the larger ephemeral river courses just after the turn-off to Kriess se Rus.

Some 20 km after Kriess se Rus, and having passed through a very flat area, outcrops of Donkerhuk granite make their first appearance. The Donkerhuk granite is a grey, biotite-muscovite post-tectonic Damaran granite, which intruded the Kuiseb schist some 520 Ma ago. After the granites, Kuiseb schist once again underlies the surface, and dark dolerite ridges occur on both sides of the road. These dolerites are of Late Jurassic age and have developed through extensional tectonics associated with the opening of the South Atlantic Ocean during the break-up of Gondwanaland. They mostly strike in a coast-parallel direction and towards the coast they become more abundant.

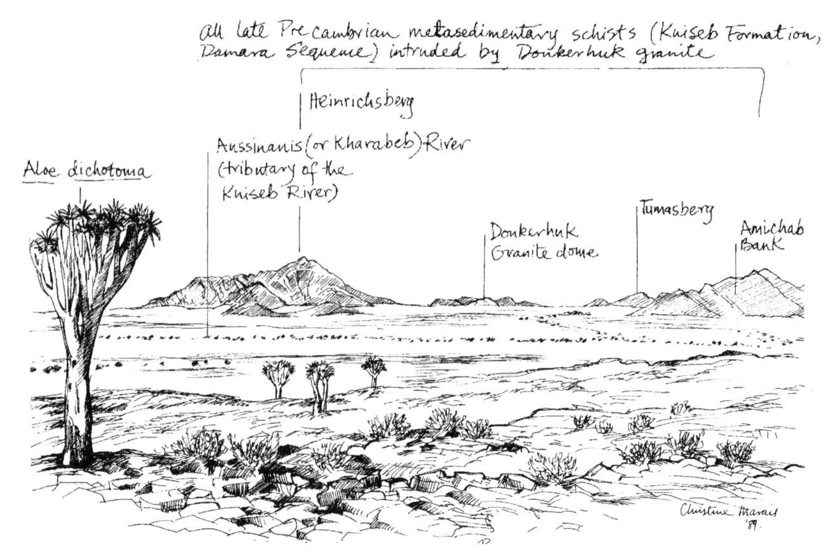


**Fig. 9.4.7:** Namib-Naukluft Park near Kriess se Rus (by Christine Marais).

As the road continues, one of the most important boundaries in the Damara Orogen, the Okahandja Lineament is crossed. This lineament represents the southern edge of the Central Zone of the Orogen, and marks major changes in stratigraphic succession, structural style and age of metamorphic events. The lineament trends NE and represents a deep penetrating zone of weakness in the crust which has repeatedly been active and which had a major influence throughout the depositional and tectonic history of the Damara Orogen (Miller 1979). However, apart from a transgression from the lighter coloured Kuiseb schists into the dark coloured schists of the Tinkas Formation, there is at first no marked change in the landscape, which is dominated by a peneplain covered with young sediments of the Namib Desert.

After some 30 km Vogelfederberg is reached, and this hill, as well as the surrounding hills consist of the pinkish-gray, foliated, porphyritic granite of the syn- to post-tectonic Salem Suite. The granite displays typical onion-skin weathering. Further on from Vogelfederberg, the road crosses meta-sediments of the Etusis Formation, Nosib Group and dolomites and mica schists of the Karibib Formation, Swakop Group, which form the ridge of the Hamilton Range to the SW of the road. Thereafter, the landscape becomes very flat, and only isolated hills of granite may be seen.

35 km after Vogelfederberg the prominent Rooikop hill, next to the Walvis Bay Airport, is reached. This hill also consists of Salem Granite. The young sediments of the surrounding peneplain are likewise underlain by this granite, which supports granite-quarrying operations that can be observed in a number of places. Further on, there is a turn-off to Rooibank, where water is abstracted from the sands of the Kuiseb River to support the coastal towns of Walvis Bay and Swakopmund, and the mining operations at the Rössing Uranium Mine E of Swakopmund. The coastal dune belt becomes visible here to the NW. It runs parallel to the coast from Walvis Bay to just S of the Swakop River. It is the only location where the Namib Sandsea extends northwards across the Kuiseb River. While floods of the Kuiseb flush out the windblown sand further to the E and thereby prevent encroachment across its course, the fact that the river usually does not reach the ocean has allowed for the establishment of this narrow coastal dune belt. Some of the highest sand dunes in Namibia, such as Dune 7, are located here. Noteworthy are the black areas on the dune surfaces, which represent wind-generated concentrations of dark heavy minerals such as magnetite, ilmenite and garnet.

Just outside Walvis Bay, the coastal flats of the Kuiseb River delta are reached. Small barchan dunes can be seen S of the road and a variety of seabirds occur in the vicinity of ponds generated by the Walvis Bay sewerage works. Some of these ponds, particularly N of the road, are dry showing a sealing sediment floor that contains appreciable amounts of gypsum. Further to the S, the coastal flats are extensively used for salt production by solar evaporation in similar, but of course much larger ponds.

Having passed the coastal dune belt, the road turns N along the coast to the W of the dune belt. There are only a few rock outcrops along the coast, although close to Swakopmund, more extensive outcrops are seen. These comprise of schist and amphibolite of the Khan Formation, Nosib Group and Kuiseb schist and calc-silicate of the

Karibib Formation, Swakop Group. Before entering Swakopmund, the road crosses the Swakop River, one of the largest ephemeral watercourses in the country.

## 9.5 Windhoek – Spreetshoogte Pass – Naukluft – Maltahöhe

This route starts in the highlands to the S of Windhoek, follows a spectacular descent of the Great Escarpment via the Spreetshoogte Pass, then borders the Namib Desert before passing the famous Naukluft Mountains to reach Maltahöhe, a total distance of 387 km (Figs. 9.5.1, 9.5.2). The first 113 km follow the same route described in 9.4, before turning off to the S on road 1265 in an area underlain by Cenozoic sediments on the farm Göllschau. The Gamsberg Mountain (see 8.10) is very prominent to the W, while to the E of the road, the Kamelberg mountain range composed of amphibolites of the Elim Formation is visible. Further to the SE, along a thrust, meta-sediments of the Elim and Gaub Valley Formations have been thrust over Gamsberg granite.

Four kilometers from the turn-off, the road enters terrane underlain by Gamsberg granite, with the hills to the east of the road composed of the same granite. There are some granite outcrops fairly close to the road, and after about 2 km, the road takes a marked turn to the W. Here, on the southern side of the road there is a large peneplain underlain by Cenozoic sediments, while some outcrops fairly close and on both sides of the road show the typical onionskin weathering of the Gamsberg granite. This landscape continues for another 2 km, before the road turns S again and eventually enters the peneplain underlain by the Cenozoic sediments. The topography in the vicinity of the road is now quite monotonous, but in the W a mountain range extending S of the Gamsberg is composed of Weener quartzdiorite which intruded meta-sediments of the Gaub Valley Formation.

Some 18 km after the turnoff, the road moves once again into terrane underlain by Gamsberg granite, and shortly thereafter by schist of the Gaub Valley Formation, characterized by large quartz segregations. The road is now approaching the Areb Shear Zone bounded by two major thrusts. Some 7 km further on, after passing some quartzite of the Sinclair Sequence, schist of the Areb Shear Zone thrust onto the quartzite forms a small ridge to the eastern side of the road. After another 5 km this mountain range is now fairly prominent on the western side of the road, but 4 km onwards the road passes the lower boundary thrust of the Areb Shear Zone, and softly undulating hills composed of schist of the Elim Formation occur on the southeastern side of the road. These hills are covered by vegetation and look distinctly different from the hills composed of schist of the Areb Shear Zone seen on the northwestern side of the road. The dark rocks of the Alberta Complex are also now showing in the distance to the W. The Alberta Complex is the most prominent of a series of ultra-mafic bodies that have intruded the Rehoboth Sequence. It covers an area of approximately 100 km$^2$ and forms an oval-shaped, intensely faulted body composed essentially of a thick succes-

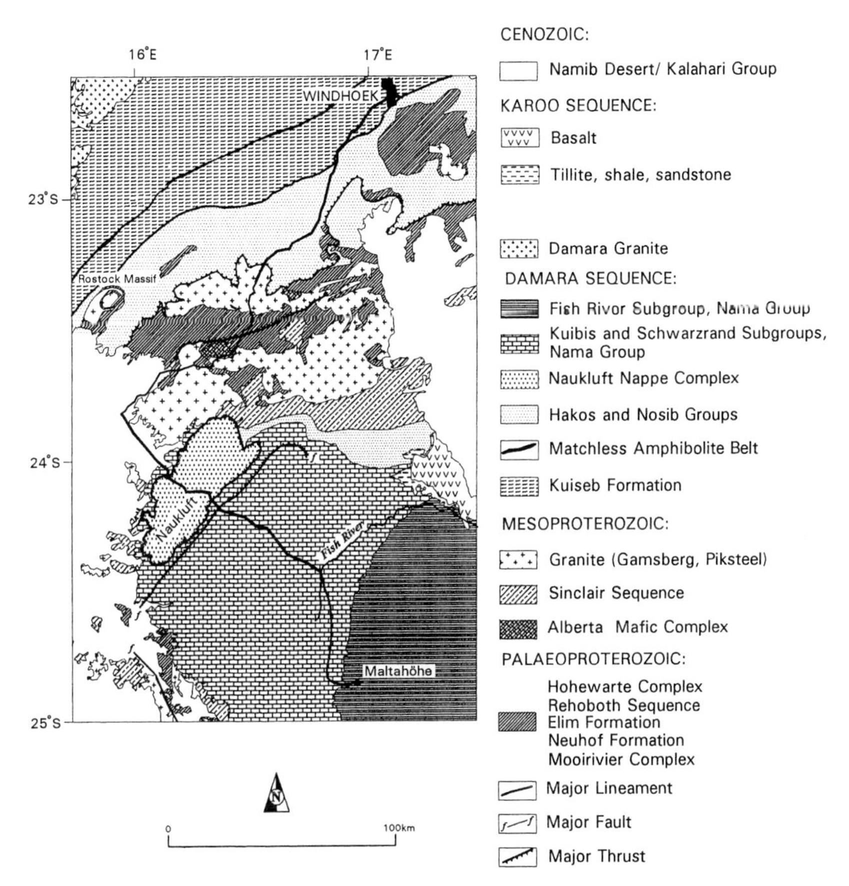


**Fig. 9.5.1:** Geological map for route 9.5.

sion of layered gabbroic rocks which was intruded at a later stage by a pegmatoid phase as well as by harzburgite and dunite (DeWaal 1966). The age of the Alberta Complex has been estimated at 1760 Ma.

The road now passes through an area underlain by Cenozoic sediments, but after 3 km rocky outcrops occur fairly close on either side of the road, with Gamsberg granite on the southeastern side and amphibolites of the Alberta Complex to the northwest. Once the farmhouse on farm Areb is reached after another 6 km, the valley has become fairly narrow with outcrops of Elim Formation schist on either side of the road. Thereafter the

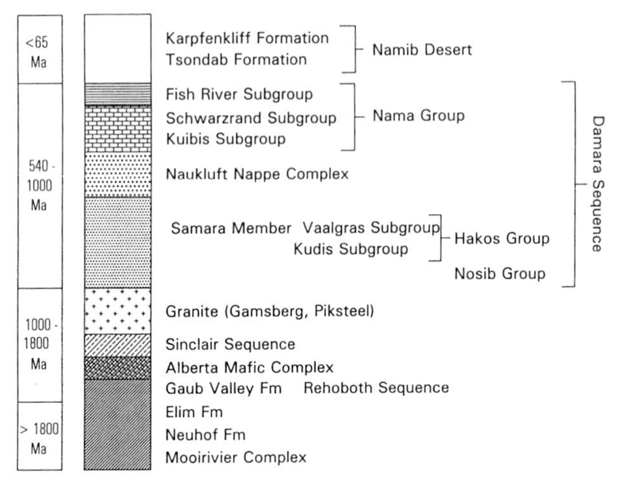


**Fig. 9.5.2:** Stratigraphic column for route 9.5.

road leads through a landscape, underlain by Gamsberg granite, and after 8 kilometers an aplitic variety of that granite forms a few smaller hills on the southern side of the road. Two kilometers further on, road D 1275 to the Spreetshoogte Pass turns off to the N in the vicinity of the old farmhouse and police station.

After the turn-off the road continues in an area underlain by granites of the Piksteel Suite, however, after just one kilometer it crosses an unconformity and once again, Gamsberg granite is the country rock. Slowly, the valley opens, and granite hills, more prominent to the N, can be seen on both sides of the road. Shortly after this the valley narrows once more, and the road winds through an undulating area underlain by Elim Formation schist. Some 9 km after the Spreetshoogte turn-off, the road moves back into granites of the Piksteel Suite, and for some 13 km until the top of the Spreetshoogte Pass, the extremely narrow valley provides numerous good outcrops (Fig. 9.5.3).

The top of Spreetshoogte Pass provides one of the most dramatic views across the Great Escarpment into the Namib Desert. The Great Escarpment here drops to the Namib plains of the Ubib basin below. The red dunes of the main Namib Sand Sea can be seen in the distance. In the middle distance, numerous pans dot the pedogenic calcrete which caps either conglomerate deposits of the Karpfenkliff Formation or sandstones of the Tsondab Formation. The Tsondab Formation (see 8.4) represents an early arid phase in the Namib Desert between 10 and 20 Ma ago, and testifies to the antiquity of the Great Escarpment. To the N-NW, extensive thrusting of the Areb Shear Zone can be observed in a hill, composed of Sinclair Sequence quartzite and rocks of the Areb Shear Zone thrust onto Gamsberg granite.

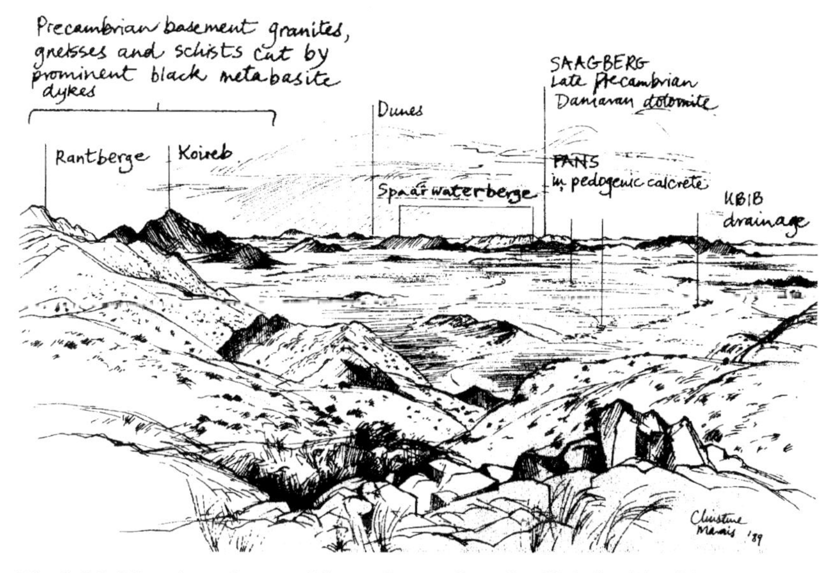


**Fig. 9.5.3:** View from the top of Spreetshoogte Pass (by Christine Marais).

The descent, one of the steepest down the Great Escarpment with a gradient of 1:4,5, begins in granites of the Piksteel Suite. These granites form the mountains along the Great Escarpment in this area. Spectacular views open up on both sides of the road, as it continues to descend, and moves into an area underlain by young Tsondab Formation sandstones, which becomes immediately noticeable in road outcrops. The bottom of Spreetshoogte, almost 1000 m below the top is reached after just 4 km.

The road now follows a fairly wide valley underlain by Cenozoic sediments. A prominent hill of buff coloured dolomites of the Kudis Subgroup, Damara Sequence, occurs right at the bottom of the pass, and other such dolomite hills occur further down the valley. They support a noticeably different vegetation with resurrection bushes and fig trees. Five kilometers after the bottom of the pass, there is a good view into the Namib Desert in the distance. The high mountains to the SE represent a continuation of the Great Escarpment composed of Piksteel Suite granite. After another 5 km, the prominent V-shaped mountains to the S of the road are made up of Gaub Valley Formation schist intruded by and thrust onto Gamsberg Granite. Some 5 km further on, prominent thrusting of Damaran marble of the Samara Member onto schist of the Gaub Valley Formation is clearly visible in a smaller hill W of the road. Shortly the road turns to the W, and the hills on the southern side of the road are dominated by schist of the Gaub Valley Formation, intensely intruded by granite and diabase dykes, which produce easily noticeable rubble fans of dark boulders. Two kilometers further on the intrusive contact between the Gamsberg granite and the Gaub Valley Formation schist can be observed close to the road. After another 6 km a diabase dyke passes within walking distance of the road, boulders of which lie right next to the road.

At this locality in the distance to the N, mountains composed of Damaran marbles, displaying intense thrusting can be seen and to the W, straight ahead of the road, the Namib Sand Sea provides a magnificent view. On the fringe of the Namib Desert, smaller hills of gneisses of the Mooirivier Complex and dolomites of the Vaalgras Subgroup, Damara Sequence, can be observed contrasting with the desert dunes.

The turn-off to the main road C14 from Walvis Bay to Maltahöhe is reached 32 km from the bottom of Spreetshoogte Pass, and the Naukluft Mountains (see 8.22) appear prominently in the south. As the route follows the road C14 towards Solitaire, the Great Escarpment composed of massive Piksteel granodiorite dominates the landscape in the E, with the Naukluft Mountains following in the SE. In the W, the vast peneplain of the Namib Desert can be seen. Solitaire is reached 9 km from the turn-off to the Spreetshoogte Pass.

Some 12 km south of Solitaire, the Noab Rivier, a major tributary to the Tsondab River is crossed, and to the E a mountain representing an outlier of the Great Escarpment and comprising of Piksteel granodiorite at the base, overlain by sediments of the Nama Group, can be observed. In the W, in the middle distance, some mountains composed of Gamsberg granite are visible and two kilometers further on, the Tsondab River is crossed. This river is a major westward drainage in the area originating E of the Naukluft Mountains and supplying water to Tsondab Vlei (see 8.28).

The road enters the Namib Naukluft Park for a short while some 15 km from Solitaire. The Great Escarpment with Piksteel granodiorite is still prominent in the E, however, as the road enters the valley of the Tsondab River, which is the major valley through the Naukluft Mountains, after about 2 km, Gamsberg Granite overlain by well stratified black limestone of the Kuibis Subgroup, Nama Group, occurs on both sides of the road. These lithologies follow the road for the next 8 km, and exhibit some spectacular intense shearing and deformation, which affected the autochthonous Nama sediments during Naukluft overthrusting (compare 8.22).

The road then turns E, and as the valley widens, the morphology of the mountains also changes. The mountain flanks now comprise green shale of the Pavian Nappe at the bottom, overlain by the gray dolomite of the Kudu Nappe of the Naukluft Complex (see 8.22). The floor of the valley itself is underlain by Cenozoic surficial deposits of sand and calcrete. After 2 km, in a valley to the SW, the contact between the two nappes can be seen halfway up the mountain. The contact is marked by the accumulation of travertine, deposited by springs developed along the contact. These form when meteoric waters percolating through the dolomite encounter the impermeable shale. The waters carry carbonate dissolved from the dolomite, which is then re-deposited, and in this way forming huge "petrified waterfalls", hence the farm name Blässkranz (compare Fig. 8.22.2).

Some 30 km from Solitaire, a valley branches off to the NE, with its flanks comprising mostly dolomites of the Kudu Nappe. Further on, after about 3 km, the valley widens considerably, and another valley branches off to the W. At the end of this valley, the limestone of the Dassie Nappe is exposed. Next to the road, which now closely follows the Tsondab River, there are abundant outcrops of calcrete. After another 5 km, the road is close to the southern fringe of a fairly wide valley, and the slopes of the mountains to the S contain a variety of lithologies, starting with green shale, boulder shale and white dolomite of the Pavian Nappe at the base, followed by grey dolomite of the Kudu Nappe on top. Massive travertine formation can also be observed. On the northern side of the valley, green shale and boulder shale of the Pavian Nappe are overlain by the massive grey dolomite of the Kudu Nappe.

As the valley slowly narrows, there is an isolated hill right next to the northern side of the road, where green shale of the Pavian Nappe and dolomite of the Dassie Nappe can be seen in outcrop. The road then takes a marked turn to the S, and limestone of the Zebra Nappe, overlain by dolomite of the Dassie Nappe form the mountains to the SW whilst in the NE, dolomite of the Dassie Nappe prevails. Intercalated into these limestones and dolomites is a layer of quartzitic shale of the Zebra Nappe, with again massive travertine development. Some 53 km from Solitaire, the valley almost closes, leaving only a small gap, through which the road leaves the Naukluft Mountains. It then closely follows the Tsondab River with ample exposures of calcrete. The mountains on either side, as well as an isolated hill in the valley are composed of dolomite of the Dassie Nappe and shale and limestone of the Kuibis Subgroup, Nama Group, which have been intensely sheared and deformed during Naukluft over-thrusting. The Büllsport farm house is reached after a further 1 km down the road.

As the road leaves the Naukluft Mountains behind, it follows a very flat peneplain with no outcrop. A few small hills can be seen in the SW and they are composed of shale with thin quartzite layers of the basal Schwarzrand Subgroup of the Nama Group. The landscape itself is underlain by Cenozoic surficial sediments, but some 13 km SE of Büllsport, the road enters terrane underlain by this shale, which is exposed in road cuts. Four kilometers further, the road descends into the small valley of the Nabaseb River, and thereby enters the catchment area of the Hardap Dam. Road cuts provide ample outcrops of shale, as well as the underlying quartzite. Intercalations of finer- and coarser-grained shale can be observed. The following ascent out of the valley also provides road cuts. To the northeast of the road, some flat-topped hills are likewise made up of sediments of the Nama Group.

For the next 20 km the road follows an undulating landscape, interrupted only by the occasional river crossing. The area is underlain by greenish Nama shale and quartzite, exposed in the occasional road cut. Some table-topped hills on both sides of the road are of the same material, capped by greenish and reddish shale. About 37 km from Büllsport, the Kaigab River is crossed, and the road ascends slightly, but still in an area underlain by the shale mentioned before. This is reflected by the darker colouring of the road, due to a difference of the material used in road building.

The road remains in terrane underlain by red shale and quartzite for the next 23 km, however, it slowly ascends to higher stratigraphic levels. Road cuts provide good outcrops in these sediments, which appear fairly massive and red, with finer grained intercalations. About 60 km from Büllsport at Nomtsas farmhouse, the crossing of the Fish River, one of the major drainages in southern Namibia, coincides with a change in the geology. The area is now underlain by reddish shale and sandstone of the upper Schwarzrand Subgroup and the descent and following ascent provide good outcrops in road cuts.

For almost the entire remaining 51 km to Maltahöhe, the road stays in terrane underlain by these reddish Nama sediments. There are river crossings of tributaries of the Fish River, which follows the road to the W, and these provide for limited relief in the otherwise monotonous landscape. Some road cuts, in which intercalations of fine-grained and coarser-grained material can be observed, do occur near these river crossings. Closer to Maltahöhe, the Schwarzrand Escarpment becomes prominently visible in the E and just outside Maltahöhe, the road descends into the valley of the Kuhab Rivier, which is underlain by Cenozoic surficial sediments. Just before the bridge over this river, immediately to the W of Maltahöhe, the road cuts through massive red sandstone of the lower Fish River Subgroup. Maltahöhe is underlain by these sediments and the town is reached 111 km from Büllsport.

## 9.6 Rostock – Solitaire – Zarishoogte Pass – Maltahöhe

The route from Rostock to Maltahöhe begins at the road junction of the C26 (Gamsberg Pass) and the C14 from Walvis Bay to Maltahöhe. It follows the edge of the Namib Desert W of the Great Escarpment, passes the Gaub Valley and the Naukluft Mountains and climbs the Great Escarpment via the Zarishoogte Pass to reach Maltahöhe, a distance of 307 km. This route is also taken to reach Sesriem and Sossusvlei (Figs 9.6.1, 9.6.2).

At the C26 turn-off, the Rostock Massif dominates the landscape to the E. It is composed of thrust slices of Palaeoproterozoic schist and amphibolite of the Gaub Valley Formation and Gamsberg Granite, whilst to the SW, in the desert, the prominent Kammberg and some lower hills comprising Kuiseb Formation schist can be seen. The road itself follows a peneplain underlain by Cenozoic sediments of the Namib Desert. After about 4 km the road passes a ridge of Tertiary Tsondab Formation sandstone, deposited some 10 to 20 Ma ago under desert conditions pre-dating the modern Namib Desert (see 8.4 and 8.21). The sandstone is frequently capped by calcrete, a common surface limestone formed under arid to semi-arid conditions through the solution and redistribution of lime by meteoric waters.

After 7 km, the Rostock Massif is now directly E of the road and the intense thrusting is clearly visible. The road continues in an area underlain by Tsondab Formation sand-

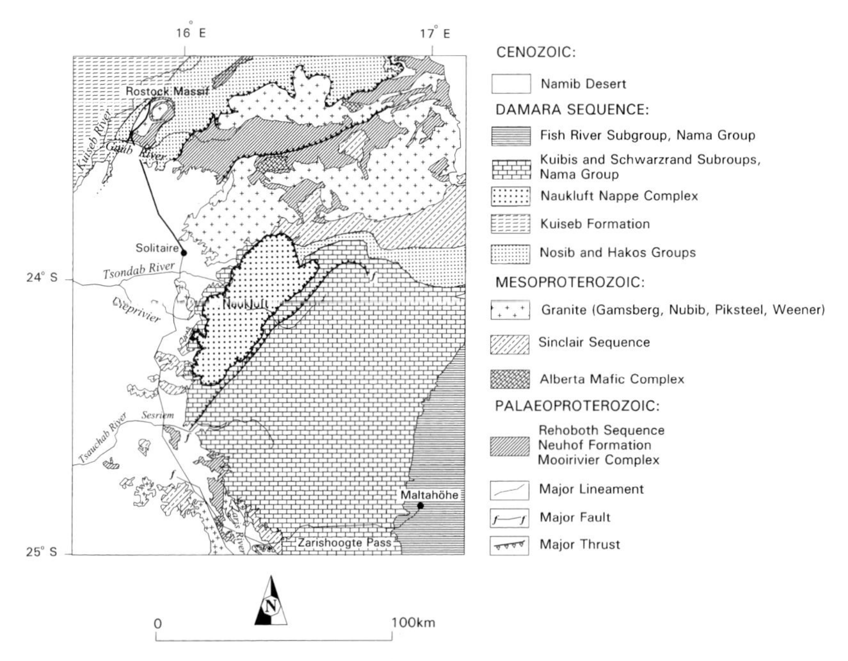


**Fig. 9.6.1:** Geological map for route 9.6.

stone and to the W one can observe quartzite and schist of the Hakos Group. The scenic descent into the Gaub Valley begins some 19 km after the C26 turn-off. The valley itself is deeply incised into schists and diamictites of the Naos Formation of the Hakos Group, and at the beginning of the descent a road cut on the western side of the road displays the Naos Formation rocks overlain by Tsondab Formation sandstone, capped by about 1 m of calcrete. These contacts can be followed as the road descends into the valley, and after crossing the Gaub River. The road winds for 2 km out of the valley, repeating the same stratigraphy seen during the descent in abundant roadcuts (Fig. 9.6.3).

Once out of the valley, the road continues in a peneplain underlain by Tsondab Formation sandstone capped by calcrete. Small rivers incised deeply into the calcrete bear witness to the sporadic flood events that can take place. Some 10 km after the Gaub River crossing, the Kammberg is passed in the W and after another 3 km, sand dunes of the Sossus Formation of the Namib Group can be seen. The dune morphology is particularly well developed on the western side of the road, and it is noticable that some dunes are covered with sparse vegetation. To the E, the Witberge, composed of light dolomite of the Hakos Group thrust onto schist of that Group become visible after

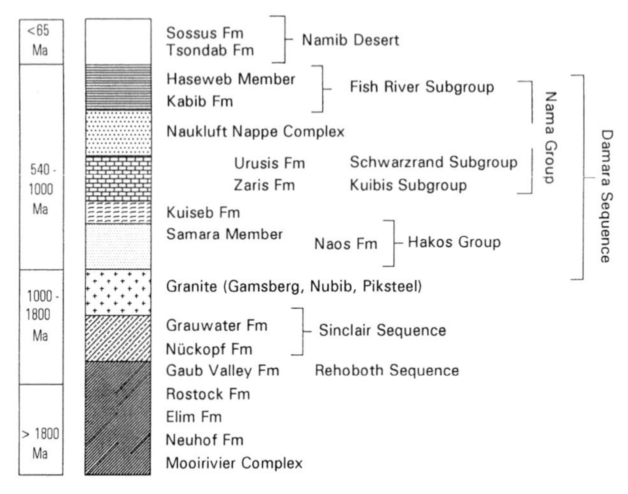


**Fig. 9.6.2:** Stratigraphic column for route 9.6.

7 km. The thrusting is well developed and easily recognisable. In the foreground, Tsondab Formation sandstone forms a hill, capped by calcrete. It is the calcrete, deposited in a pan, which is responsible for the flat-topped appearance, and is therefore limited in extend.

Elongate mountain ranges on either side of the road dominate the landscape for the next 20 km. In the W, the prominent Saagberg is composed of light dolomite and marble of the Samara Member of the Hakos Group. The contact of Gaub Valley Formation schist and the dolomites and marbles, which have been thrust onto it, can be seen in some isolated outcrops closer to the road. To the E, the mountains display a spectacular thrust separating light coloured Samara Member marbles and dolomites from the underlying darker Piksteel Suite granites and Mesoproterozoic amphibolites and closer to the road some Piksteel Suite granites occur. Having passed these mountains, the peneplain opens up with only a few smaller hills visible in the W. These hills are composed of Samara Member dolomite and marble and Palaeoproterozoic gneiss of the Mooirivier Metamorphic Complex. The turn-off to road D1275 (Spreetshoogte Pass) is reached 45 km after the Gaub River.

As the route follows the road C14 towards Solitaire, the Great Escarpment, here composed of massive Piksteel granodiorite, dominates the landscape in the E, with the Naukluft Mountains (see 8.22) following in the SE. In the W, the vast peneplain of the Namib Desert can be seen. Solitaire is reached 9 km from the turn-off to Spreetshoogte Pass. At Solitaire, the route turns along road 36 straight to the S. The smaller mountain

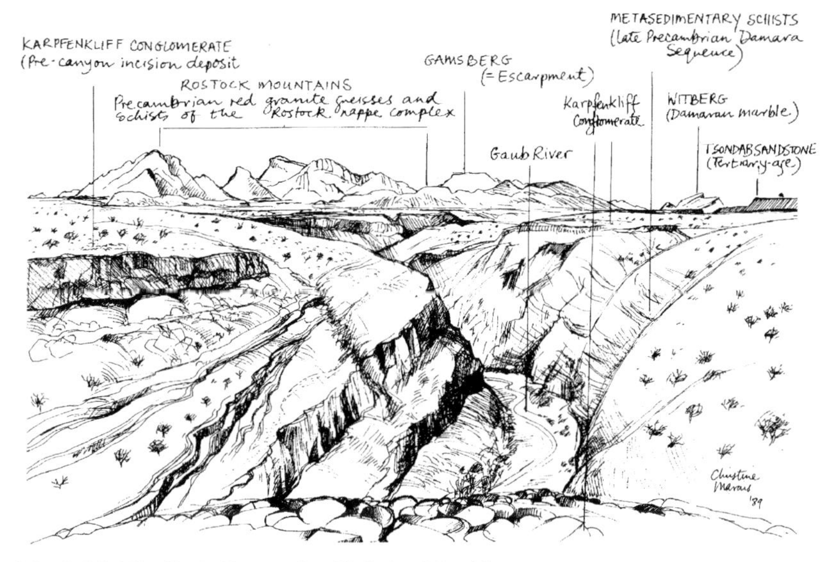


**Fig. 9.6.3:** The Gaub Canyon (by Christine Marais).

range ahead of the road is again composed of Piksteel Granodiorite and sediments with intercalated volcanics of the Sinclair Sequence. To the W, the Namib Desert contains some inselbergs of amphibolite of the Elim Formation.

About 7 km S of Solitaire, just before the crossing of the Tsondab River, the Naukluft Mountains now appear prominently to the E (Fig. 9.6.4), with layered Nama sediments overlying Piksteel Granodiorite basement. Further to the SE, the contacts with the nappes of the Naukluft Complex can be seen and in the distance in the W, there are some hills composed of marble of the Samara Member, Hakos Group, of the Damara Sequence. After another 3 km, the mountains that appear ahead of the road consist of granites of the Gamsberg Suite. As the road approaches these mountains, it gains some elevation, providing a good view into the Namib to the W. Inselbergs of the same basement rocks are prominent in the foreground, while the Namib Sand Sea can be seen in the distance. Some 15 km S of Solitaire, the road passes the Gamsberg Granite mountains in the E, and the intercalated Elim Formation amphibolites become clearly visible.

About 8 km further on, a hill in the W has granite and amphibolite at the base, with stratified Nama sediments on top. Once this hill is passed and after another kilometer, there is a magnificent view into the Namib Desert with its partly vegetated, red sand dunes. Just below the red sand dunes, the Tsondab Sandstone cliffs at Dieprivier

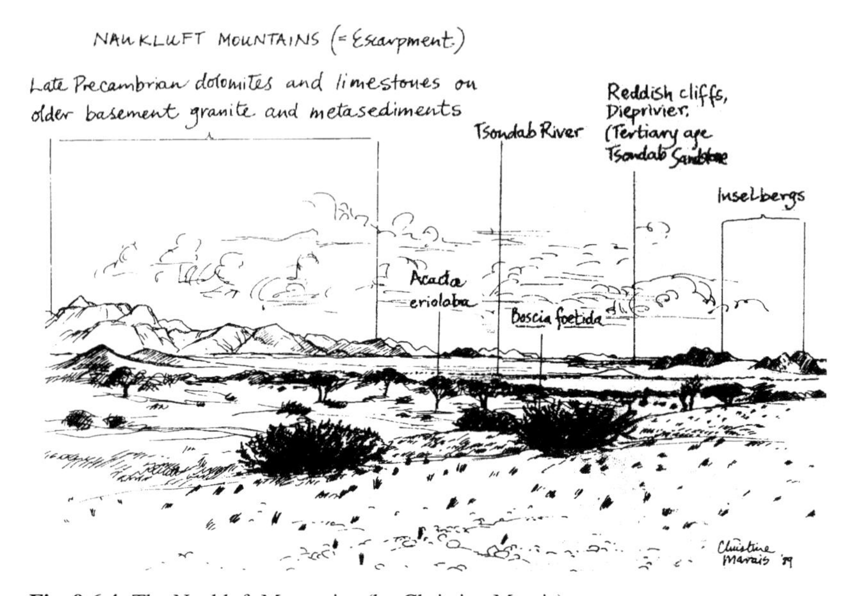


**Fig. 9.6.4:** The Naukluft Mountains (by Christine Marais).

(see 8.4) become visible, and clearly show the relationship of these ancient desert deposits with the modern Namib Sandsea (Fig. 9.6.5). Further to the W an isolated inselberg is composed of Gamsberg granite and gneiss of the Neuhof Formation. To the SE the high mountains of the Naukluft, with quartzite of the Grauwater Formation, Sinclair Sequence at their base, become visible once more in the Great Escarpment.

Several courses of the Dieprivier are crossed between 30 and 33 km S of Solitaire. This river, which has carved the cliffs of Tsondab Sandstone on the farm Dieprivier, is very wide, and has incised its bed into massive deposits of calcrete, which are seen on the sides of the channels. Its waters feed Tsondabvlei in the desert to the W (see 8.28), whilst to the E, the Naukluft Mountains remain prominent. The smaller hills seen on both sides of the road, approximately 40 km S of Solitaire, are composed of Piksteel Granodiorite topped by Nama sediments. The entrance to the Namib Naukluft Park is reached some 2 km from this point.

About 4 km into the park, some hills with folded meta-volcanics of the Nückopf Formation, Sinclair Sequence appear E of the road. The road essentially follows the foot of these hills, and layers of dark and light meta-volcanics can be clearly seen. The Namib sand dunes are now coming closer in the W, and here they are well vegetated. The mountains seen straight ahead are again part of the Naukluft Complex, however, they also expose the basement to the Naukluft nappes and contain some Nückopf Formation

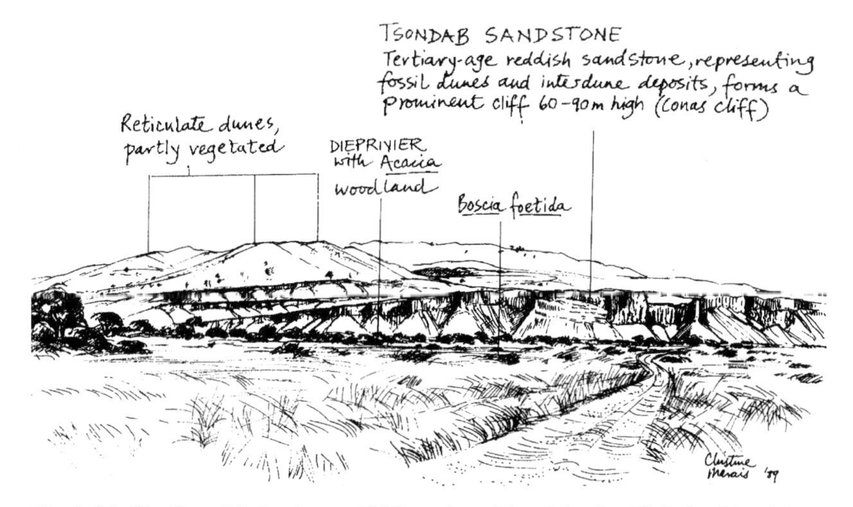


**Fig. 9.6.5:** The Tsondab Sandstone Cliffs on farm Dieprivier (by Christine Marais).

rocks. The landscape in the E is now dominated by smallish hills of the Sinclair Sequence, and the actual Great Escarpment is no longer visible.

Similarly, some 10 km after the entrance to the Park, the Namib Desert is no longer visible in the W, obscured by mountains composed of black limestone of the Kuibis Subgroup, Nama Group. The strata are dipping chiefly to the S, although there is also some folding. The folding and layering is clearly visible and enhanced by the sand dunes, which have accumulated on the eastern slope of the mountains. As the road turns once more SE, a massive mountain range composed of Nubib granite and gabbro appears ahead, whilst directly to the E meta-sediments of the Grauwater Formation are passed.

Some 60 km south of Solitaire, the Naukluft Mountains, representing the Great Escarpment, become fully visible once more. Three kilometres further on, the massif composed of Nubib Granite and gabbro is seen in the W and also extends slightly to the E of the road. Thereafter, the road leaves the Namib Naukluft Park and towards the S, a mountain range of Gamsberg Granite extends in an E-W direction well into the desert. After another 4 km, the western extension of the Naukluft Mountains is passed in the E, revealing a view to the SE onto hills of gneiss of the Mooirivier Complex in the foreground, with stratified Nama Group mountains in the distance. The road then passes the mountain range of Gamsberg granite in the W and descends into the wide valley of the Tsauchab.

Some 72 km S of Solitaire, there is the turn-off to Sesriem (see 8.27) and Sossusvlei (see 8.28). Some of the highest sand dunes in the World are visible in the distance in the

W, while the Great Escarpment mountains in the E expose stratified grey and black limestone of the Kuibis Subgroup, Nama Group, overlying gneisses and granites of the basement. These basement rocks also form the high mountains seen to the S of here. After 6 km, a large river, which joins up with the Tsauchab further W, is passed, and the Tsauchab itself, which provides the water to Sossusvlei (see 8.28), is crossed 10 km S of the turn-off to Sesriem and Sossusvlei. In this area the limestone mountains of the Nama Group are still visible in the distance in the E, whilst the immediate vicinity on both sides of the road is dominated by mountains composed of granite of the Gamsberg Suite and Nubib Granite.

After another 2 km, a major tectonic feature, the Hebron fault is crossed, and is visible right next to the road. This feature is one of the most impressive young scarps in western Namibia (Fig. 9.6.6). It is a sub-vertical fault with the western block downthrown by up to 7 m and can be traced for more than 35 km in a SE-NW direction. The fault is locally developed in calcrete, and the involvement of these Cenozoic sediments indicates a recent age for the faulting possibly within the last 100 000 years. The fault is interpreted as an incipient stage of continental rifting, possibly caused by the rejuvenation of older structures. Water rises on the fault plane, and boreholes in the vicinity of the fault tap warm water at a depth of 70 m. The location of the farmhouse on Hebron, after which the fault takes its name, right next to the fault is based on the presence of water in the area.

**Fig. 9.6.6:** The Hebron Fault (photograph by Branko Corner).

Just after passing the Hebron Fault, basement granites are now adjacent to the road in the W, and the view into the Namib Desert is blocked by hills composed of basement rocks. Towards the SW, the Nubibberge, after which the Nubib Granite is named, become visible, while towards the SE, the lower hills are composed of metamorphic rocks of the Neuhof Formation. The first glimpse of the Zaris Mountains appears some 20 km S of the turn-off to Sesriem and Sossusvlei. The Zaris Mountains form the Great Escarpment in the area and are composed of basement gneisses, granites and schists of the Neuhof Formation, Gamsberg Suite and Sinclair Sequence, overlain by flat-lying black limestones of the Nama Group. The flat-lying limestones are responsible for the "table mountain" morphology of the Zaris Mountains, which soon dominates the view to the SE.

The conspicuous Rooikop hill, composed of red metamorphic rocks of the Neuhof Formation, is passed directly E of the road about 37 km S of the Sesriem turn-off, while the Nubib Mountains are still prominent in the SW. As the road progresses towards the S, the valley narrows, and the basement hills at the foot of the Zaris Mountains now obscure the view of the Nama Group sediments on top. As the road passes the farm Neuhof, it now runs in a narrow valley bounded by basement, forming smooth, undulating hills on either side. About 20 km S of Rooikop, on the farm Hammerstein, the valley widens once more, and the Zaris Mountains re-appear in the E. To the SW, the flat-topped Hammerstein Mountain, giving the farm its name, is visible and is composed of Neuhof Formation metamorphics at the base overlain by Nama Group limestones, and represents on outlier of the Zaris Mountains to the W of the Great Escarpment.

As the road approaches the bottom of Zarishoogte Pass, the Zaris Mountains dominate the view to the E. It is noteworthy that in some places the sedimentary cover has already been eroded, and that further erosion is now producing well developed onionskin weathering of the basement rocks. One such hill occurs directly E of the road about 33 km south of Rooikop, and is named Alwynkop. Thereafter, the "table mountain" morphology dominates the landscape on both sides of the valley, and after another 10 km, the Zaris River with its big Kamelthorn trees and abundant Sociable Weaver nests, is crossed. The road then climbs within the valley of the Zaris River, and while there are still some outcrops of basement rocks W of the road, the view ahead presents limestone of the Kuibis Subgroup, Nama Group. Four kilometres after the river crossing, the road passes a major thrust, along which granites of the Gamsberg Suite were thrust over red quartzites and conglomerates of the Aubures Formation, Sinclair Sequence.

The signpost "Zarishoogte Pass" is passed some 10 km after the Zaris River, and after a very steep ascent of only 2 km the top of the pass is reached. During the ascent, the road passes the overlying grey and black limestone of the Zaris Formation, Nama Group. Once the route has reached the top of the Great Escarpment, it follows a wide, gently undulating peneplain underlain by the same black limestones that were encountered in road cuts during the ascent up Zarishoogte Pass. Abundant limestone boulders are strewn all over the peneplain, and in the distance, the Schwarzrand Plateau (see 9.8) appears in the east. The limestone can also be examined in a deeply incised riverbed NW of the road, 4 km from the top of the pass, and also in roadcuts some 7 km further on.

As the road approaches Maltahöhe, the Schwarzrand Plateau becomes more prominent in the E, however, the landscape, remains very flat, with no outcrop. Some 50 km from the top of the Zarishoogte Pass, the road joins the main C14 road from Maltahöhe to Helmeringhausen. The road now follows the fairly wide valley of the Kaseweb River, which is underlain by Schwarzrand sediments, although some red sand dunes of Cenozoic age can also be seen in the valley. Outcrops of greenish shale and sandstone of the Urusis Formation, Schwarzrand Subgroup, can be observed in road cuts some 12 km before Maltahöhe. Just outside Maltahöhe, the road ascends the stratigraphy and reaches the sandstone and shale of the Kabib Member, which unconformably overlies the rocks of the Schwarzrand Subgroup. The route then crosses the Kuhab River valley and reaches Maltahöhe some 231 km from Solitaire. Just before the bridge over the Kuhab River, the road cuts through massive red, cross-bedded sandstone and mud-cracked shale of the Haseweb Member, lower Fish River Subgroup. The town of Maltahöhe and environs is underlain by these sediments.

## 9.7 Windhoek – Rehoboth – Mariental

by Thomas Becker

The 270 km long section of national road B1 between Windhoek and Mariental passes through a variety of rock types covering almost every chapter of Namibian earth history from the Palaeoproterozoic Hohewarte Complex, through the Mesoproterozoic Sinclair and Rehoboth Sequences and the Neoproterozoic Damara Sequence to the Palaeozoic to Mesozoic Karoo Sequence and Tertiary phonolites, as well as Kalahari calcrete and sand cover (Figs 9.7.1, 9.7.2).

Leaving Windhoek and heading towards Rehoboth, the Geological Survey of Namibia is located in Aviation Road near Eros Airport and opposite the Safari Hotel. Geological maps covering the country at various scales and a wealth of publications are available for purchase. Of special interest are the Geological Survey library and museum.

In the vicinity of the Geological Survey, the Windhoek valley is underlain by Cenozoic sediments deposited in the Windhoek Graben, which in turn are underlain by schist of the Kuiseb Formation of the Damara Sequence. To the SW and the SE, the Auas Mountains mark the southern boundary of the Windhoek valley. They were formed by intense thrusting during the Damaran Orogeny and chiefly consist of Auas Formation quartzite of the Hakos Group, Damara Sequence.

The first bedrock is found in a road cut some 6 km S of Windhoek, close to the bridge where the road crosses the railway line. It consists of quartzite of the Auas Formation interbedded with schist. For the next few kilometers the road runs through this unit, showing locally minor intercalations of marble and magnetite quartzite. The approximately 730 Ma old marine sediments of the Auas Formation were deposited in a spreading ocean on the continental shelf of a passive margin. Sedimentary structures

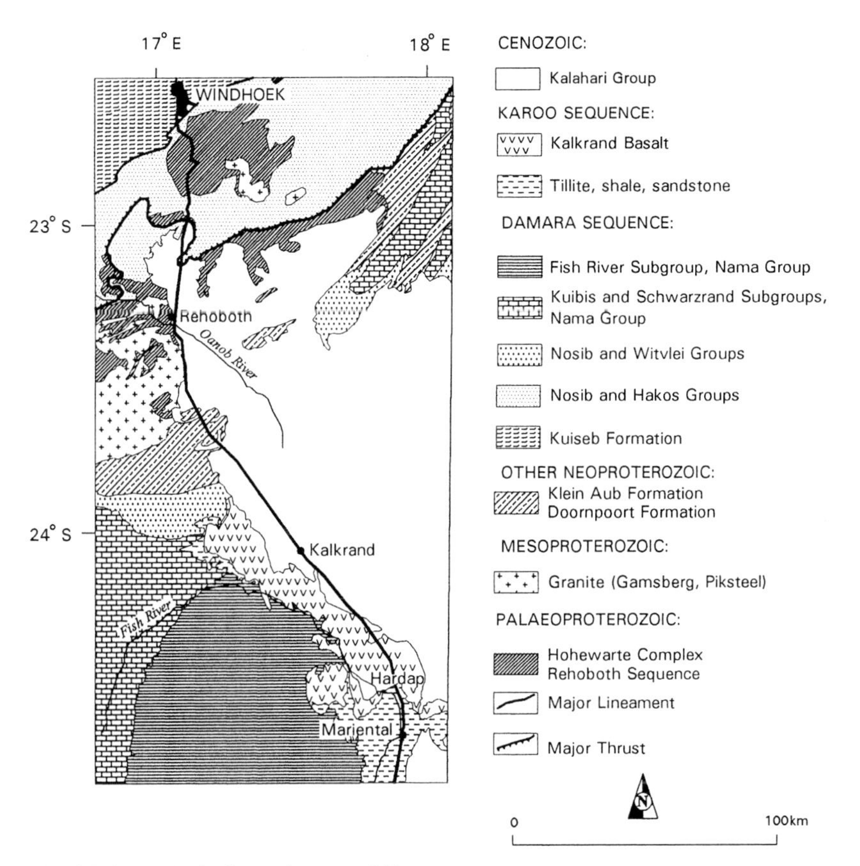


**Fig. 9.7.1:** Geological map for route 9.7.

were almost completely destroyed by subsequent metamorphism during the Damaran Orogeny. Intensely folded rocks were later thrust towards the SW, which resulted in the formation of the impressive mountain range of the Auas Mountains. Originally horizontal layers were tilted and today dip at shallow angles to the NW. These mountains are among the highest in the country, with Moltkeblick at 2479 m above sea level being the second highest peak in Namibia. At their ridge-like summits, the quartzites, which are resistant to weathering, terminate in steep, south-facing cliffs.

Some 8 km S of Windhoek, the road reaches the base of the Grossherzog Mountain, which is marked by a radio tower, and for the next 5 km the winding road rises through

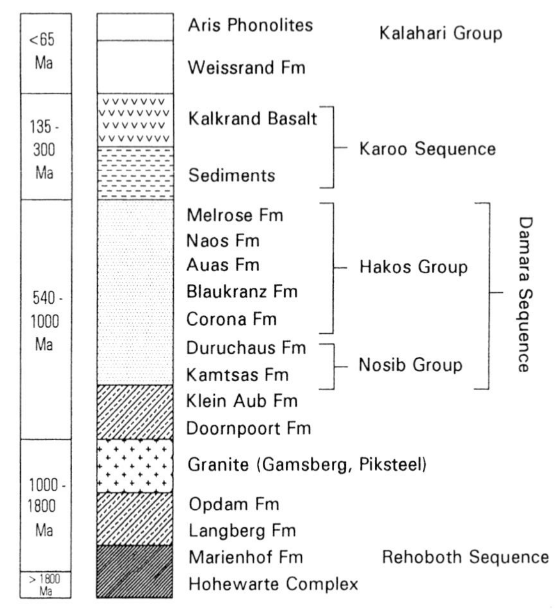


**Fig. 9.7.2:** Stratigraphic column for route 9.7.

a valley to a 1900 m high pass. This valley is a surface expression of the prominent fault system, which formed the Windhoek Graben in the Mid-Tertiary, some 35 Ma ago. The N-S trending Windhoek Graben is a tectonic structure of more than 150 km length and some 20 km width. It formed as a result of major extensional forces that occurred in the Mid-Jurassic some 170 Ma ago, and heralded the break-up of the continents prior to the formation of the South Atlantic. To accommodate crustal stress, individual blocks were displaced vertically with only minor horizontal movement along moderately dipping faults. At the top of the pass, some 16 km S of Windhoek, there is a good panorama of the cliffs of the Auas Mountains towering above a flat landscape underlain by the Hohewarte Complex, interspersed with low hills of granitic intrusions.

Block faulting is often accompanied by volcanism, with hot mantle magma rising to the Earth's surface along deep-reaching faults, which serve as passageways. In the southern extensions of the Windhoek Graben, mainly trachytes and phonolites have been emplaced, and for the next 15 km a Mid-Tertiary volcanic field extends along the road. The volcanic rocks form prominent hills, up to 200 m high, such as the Schildkrötenberg and Huquanis phonolites N of Aris, or plugs like the Gocheganas trachyte, about

10 km S of Aris. At Aris, the phonolites are quarried intermittently for railroad ballast at two locations, and the easily accessible quarries, about 1 km S of Aris, provide an excellent opportunity to study contacts, internal structures and mineralogy. In fresh hand specimen the phonolite is grayish green and contains large crystals within a fine-grained groundmass. The main components of the alkali-rich rock are aegirine, sanidine, nepheline and natrolite, with zeolites filling miarolitic caves and vugs. Apart from these main components, a variety of secondary, and very exotic minerals also occur as very small crystals (see 4.).

The quartz-feldspar gneisses inter-layered with meta-carbonates and amphibolites of the Hohewarte Complex, which surround the volcanic rocks, represent one of the most ancient crustal domains in Namibia, being more than 1800 Ma old. They form a circular domal structure with a diameter of approximately 40 km, and in turn are enveloped and locally covered by much younger rocks of Damaran age. Good outcrops of the Hohewarte Complex rocks occur in a road cut 1 km S of Aris.

Before the intrusion of the comparatively young phonolites and trachytes, granitic magmas invaded the Complex during the Damaran and Kibaran Orogenies, some 500 and 1200 Ma ago respectively. An unusual example of this earlier magmatism is encountered some 2 km N of Aris, to the E of the road, where, a Damaran pegmatite ridge known as the Falcon Rock is composed entirely of large feldspar crystals and minor muscovite.

The Gocheganas Hill ("Backenzahn"), some 10 km S of Aris, with its dragon-tooth top of trachyte, is an outstanding landmark E of the road. It is another representative of Mid-Tertiary volcanism, and nearby, the metamorphic rocks of the Hohewarte Complex have been silicified and hardened by hydrothermal waters. These waters migrated from the surface downwards, preferably on faults, and were heated up by the ascending magma, and to this day, hot springs bear witness to the increased heat flow in this zone. In the distance in the W the prominent Bismarckfelsen and other hills are composed of erosion-resistant quartzite of the Melrose Formation.

As the road continues towards Rehoboth, the Hohewarte Complex becomes covered by the glaciogenic Naos Formation, Damara Sequence, an equivalent of the Chuos Formation in the Karibib-Usakos area. Rocks of the Naos Formation can be seen in the mountains 40 km N of Rehoboth, to the W of the road. During the Damaran Orogeny they were thrust some considerable distance to the SW, onto the Hohewarte Complex, where they now form a tectonic nappe. The mountains in the distance in the E are composed of rocks of the Nosib and Hakos Groups of the Damara Sequence, likewise thrust onto the Hohewarte Complex.

Some 51 km south of Windhoek, the road passes the turn-off to the defunct Oamites Copper Mine. This mine was operational at the turn of the last century and again between 1971 and 1984. The stratiform, disseminated mineralisation is hosted by quartzites of the Oamites Formation, part of the Hohewarte Complex. Bornite and chalcopyrite are the main ore minerals.

S of the turn-off to Groot Aub, another quartzite ridge occurs in the W. It is the southernmost relic of a formerly continuous quartzite layer of the Melrose Formation, Damara Sequence, found some 50 km to the N. It conformably overlies the Naos Formation.

After passing this ridge, the Naos Formation is left behind and the road descends into subsequently lower levels of Damaran stratigraphy. First, graphitic schist, quartz-mica schist and marble of the Blaukrans Formation outcrop at the foot of some prominent hills composed of the white to bluish marbles of the Corona Formation. These rocks indicate a change in depositional environment from calm deep water with anoxic conditions (Blaukrans Formation) to a marine transgression onto a former continent (Corona Formation). Continuous crustal stretching caused the latter, and resulted in the formation of several E-W trending half grabens which gradually sagged below sea level.

Still further S, about 25 km N of Rehoboth, bluish-grey rocks occur on both sides of the road. These are the shale, siltstone, micaceous sandstone, arkose, conglomerate and minor limestone of the Duruchaus Formation, Nosib Group, which were laid down in lakes and rivers bounded by steep hills. This depositional setting marked the onset of Damaran rifting and is comparable to the East African Rift Valley of today. The heterogenous, slightly deformed sediments occupy a broad synclinal structure, and to the S they inter-finger with the fluvial quartzites and conglomerates of the Kamtsas Formation. Ranges formed by these oldest Damaran rocks can be seen E of the road. They unconformably overlie the Rehoboth Sequence, which already had undergone deformation and erosion for more than 300 Ma before the Kamtsas Formation was deposited some 850 Ma ago.

The contact between the Rehoboth Sequence, the intrusive Gamsberg Granite and the Damara Sequence lies some 10 km N of Rehoboth on the fringe of a vast peneplain. Unfortunately, most of the plain is covered by thick Cenozoic sediments, and only a few inselbergs along the road provide evidence of this older crust. The first outcrop, a granitic inselberg E of the road is encountered some 6 km N of Rehoboth.

As Rehoboth comes into view, a prominent hill W of the town is topped by a radio tower. The hill consists of mafic schist and meta-basalt of the Opdam Formation, which is probably some 1200 Ma old. Pillow structures occur at one locality and testify to subaquatic extrusion of the lavas. The formation is crosscut by several prominent diabase dykes showing spheroidal weathering. Beyond Rehoboth, the meta-basalts grade into, and are probably underlain by, felsic volcanics, which represent former hot rhyolitic ash flows and pyroclastic breccias. Their high silica content make them resistant to weathering. They form a long prominent ridge, the Langberg, which extends for more than 30 km to the E of Rehoboth. Quartzites and arkoses interbedded with the volcanic rocks locally contain primary sedimentary structures such as cross-bedding and grading. Good examples are found along the Oanob River. This river has been dammed upstream at Oanob Dam to secure the water supply of Rehoboth. At RehoSpa, a number of pools are fed by local hot springs, and like at Windhoek and Gross Barmen, the regional N-S trending fault system facilitates the circulation and heating of meteoric water in deeper layers of the Earth's crust.

On leaving Rehoboth, the basal unit of the Langberg Formation can be examined in a 300 m long road cut, just S of the bridge across the Oanob River. The main rock type is an unsorted conglomerate, in a fine-grained chlorite-phyllite matrix. The pebbles are of diverse origin, ranging in size from a few centimeters to more than one meter, deposited on the steep slopes of the initial rift by debris flows, forming submarine fans. Intercalated with the conglomerates are several porphyry sills covered and underlain by dark metamorphosed tuff. The Langberg ridge itself is prominently visible towards the E.

As one progresses southwards, the debris flow deposits rest on green and red schist of the Marienhof Formation, which forms part of the Rehoboth Sequence. Crossing this contact, 600 Ma of Earth's history are passed, and the only remainder of this immense time span is a 100 m thick discontinuous layer of pure white quartzite which caps the schist. Many of the pebbles of the Langberg conglomerates were derived from the quartzite and as it overlies the schists with a sedimentary unconformity, it must be younger than the Marienhof Formation, indicating a hiatus.

The Swartmodder Mine, can be seen to the W on a hill some 6 km S of Rehoboth. This dormant gold and copper mine is hosted within schist, micaceous quartzite and meta-lava of the Marienhof Formation, intruded by granite. Copper in the form of chalcopyrite, malachite, chrysocolla and chalcocite occurs mainly in the sheared lavas and schist forming large rafts within the granite. Altogether more than 165 tons of high-grade, hand-sorted concentrate have been produced over the years at Swartmodder, and intensive investigations between 1972 and 1981established 314 000 tons of ore reserves grading 2,7 % copper.

Further S, the country becomes flat in the E, while to the W granodiorite of the Piksteel Intrusive Suite forms low hills. The Piksteel Suite is associated with the evolution of the Rehoboth Sequence, and in common with the other units of this Sequence it exhibits a gneissic fabric absent in younger rocks. Piksteel intrusives represent an extremely large sill of batholitic dimensions, covering an area of more than 1000 $km^2$. Further S these rocks are intruded by the more balloon-shaped granitoids of the Gamsberg Suite.

The Tropic of Capricorn is passed some 20 km S of Rehoboth, and before leaving the Rehoboth geological province, one more mountain ridge is negotiated. This ridge is formed by the rocks of the Doornpoort Formation, resting unconformably on the Gamsberg Granite Suite, which at the onset of Doornpoort times was exposed by erosion. Conglomerates and unsorted talus breccias, which form the lowermost deposits of the Doornpoort Formation filled hollows in the old, uneven land surface, and subsequent layers of sediment gradually buried the granite hills. Some hills of Doornpoort Formation rocks can be seen E of the Awaseb River crossing, which is reached some 28 km S of Rehoboth.

Apart from the basal coarse clastic sediments, the 4500 m thick Doornpoort Formation consists mainly of a monotonous well-bedded fine-grained quartzite and slate, with local lenses and layers of purplish limestone, although minor rhyolitic tuffite, pyroclastic rock and amygdaloidal basalt occur near the base. The colour of the quartzite layers varies from purplish to grayish red, depending on the feldspar content, with the feldspar-rich rocks displaying a white speckling. The thickness of the individual layers increases towards the top, from a few tens of centimeters to a few meters. This red-bed succession contains a number of well-preserved primary sedimentary structures. These include horizontal lamination due to grain size variation, cross bedding, ripple marks, mud cracks and flute casts, pointing to a shallow water to terrestrial depositional environment.

Some 39 km S of Rehoboth, a radio tower marks a hill composed of Doornpoort Formation quartzite and some 3 km further on, the same quartzite can be observed in a road cut. As the road continues, more quartzite hills, with minor intercalated rhyolite, occur on both sides of the road. Doornpoort lithologies are exposed in another road cut at the Tsumis River crossing.

Leaving this last important outcrop of Mesoproterozoic basement behind, the landscape changes into a vast, monotonous Mesozoic plain, which is part of the Kalahari basin. Large parts are covered by the calcrete-cemented conglomerate and coarse scree of the Weissrand Formation, Kalahari Group, of Lower Tertiary age, which dips slightly to the NE where it disappears under red Kalahari dunes. The prominent dune field is characterized by NW-SE trending parallel dunes indicating the predominant wind direction during the last glacial period some 16 000 to 20 000 years ago. At this time a large high-pressure cell was circulating over the subcontinent and the resulting strong, cold, dry winds produced the dunes. Just before Kalkrand, on the eastern side of the road, there is a major calcrete-floored pan, which is a typical landform of the Kalahari Basin.

S of Kalkrand the dune field extends along the eastern side of the road for more than 30 km before encountering the first outcrops of Kalkrand Basalt. These Early Jurassic basalts, approximately 180 Ma old, accompany the road for a part of the rest of the journey to Mariental.

The Kalkrand basalts, which cover an area of approximately 10 000 $km^2$ to the NW of Mariental, half of which is covered by Kalahari Group sediments, overlie Triassic and Late Carboniferous to Early Permian Karoo sediments in the E, while overstepping Neoproterozoic to Early Cambrian rocks of the Nama Group in the W. The eruption of these basalts resulted from the early break-up of Gondwanaland, when southern Africa started to separate from the landmasses in the E. S of the outcrop area, the three major basalt units, separated by two thin sediment layers, have an overall thickness of approximately 100 m, whereas up to 300 m are reached in the NE, with the best exposures centered around the Hardap Dam. Most interesting are the inter-bedded fluvio-lacustrine horizons, which preserve a record of the complex interaction between Karoo volcanism and contemporaneous sedimentation. Interesting outcrops can be seen as the main road descends towards the turn-off to Hardap Dam, and also along the road leading to the dam. Hardap Dam is the largest dam in Namibia, with a dam wall built by utilizing a gorge of the Fish River in the Kalkrand basalts. The water is extensively used for irrigation, and also for freshwater fisheries.

Shortly before the dam turn-off, the road has reached a lower stratigraphic level, and runs in terrane underlain by sediments of the Dwyka Group of the Karoo Sequence, and the road remains in Dwyka Group terrane until Mariental. There is little or no outcrop, however, and the peneplain is used extensively for crop irrigation. Just before Mariental is reached, some 267 km S of Windhoek, the Weissrand Plateau can be seen in the distance in the E.

## 9.8 Mariental – Maltahöhe – Helmeringhausen – Aus

Leaving the main B1 road shortly after Mariental, this 365 km long route starts off in rocks of the basal Karoo Sequence. It then progressively moves into the lower stratigraphic levels of the Nama Group, thereby providing a complete profile through this Neoproterozoic to Early Cambrian sequence, before it continues in Mesoproterozoic terrane partly covered by Cenozoic sediments (Figs 9.8.1, 9.9.1).

A thin layer of Cenozoic alluvial sediments of the Fish River covers the sediments of the Permo-Carboniferous Dwyka Formation in the area of the C19 turn-off, and after 1 km the Fish River is actually crossed. The Fish River is the major ephemeral watercourse in southern Namibia, and is a tributary of the Orange River, which dewaters in-

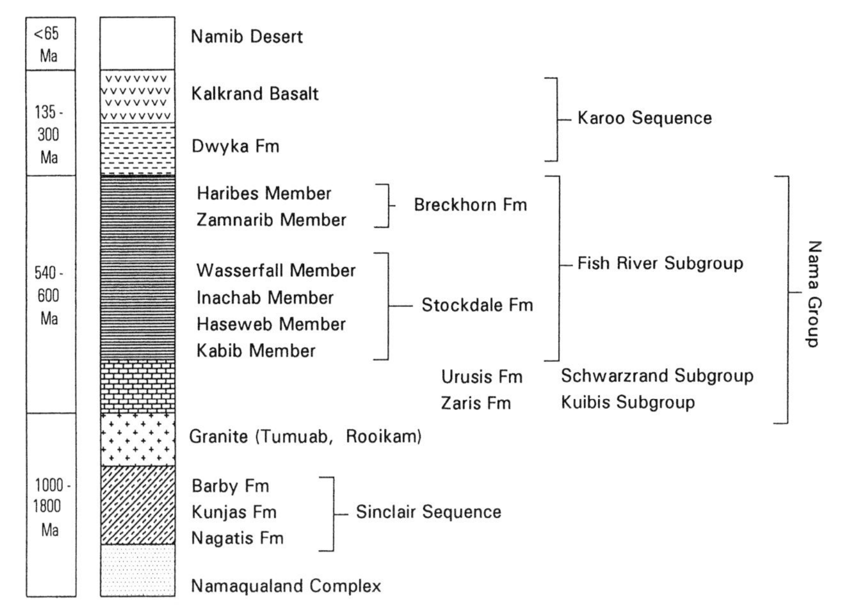


**Fig. 9.8.1:** Stratigraphic column for route 9.8.

to the Atlantic Ocean. The road continues in gently undulating landscape underlain by Dwyka Formation sediments for the next 25 km. The Dwyka Formation was deposited under glacial conditions some 300 Ma ago, with a basal tillite overlain by mudstone, sandstone and sandy limestone. Some small outcrops of the overlying Kalkrand Basalt, which formed at the end of the Karoo period during Jurassic times, can be seen soon after the C19 turn-off. After some 11 km, basalt mountains can be seen to the N of the road, and are part of the Kalkrand Basalts surrounding Hardap Dam. Outcrops seen closer to the road, including a small hill 18 km from the turn-off, are composed of Dwyka sediments.

Some 25 km from the C19 turn-off, the first outcrops of rocks of the Nama Group can be seen. The Nama Group covers large parts of southern Namibia and was deposited in a shallow marine environment during the late stages of the Damaran Orogeny as a molasse derived from the erosion of the uplifted Damara and Gariep belts. The rocks in the outcrops next to the road are red sandstones of the Haribes Member of the upper Fish River Subgroup. The landscape is still gently undulating, but a change in the colour of the soils to shades of red is immediately noticeable. As the road approaches the Lewerrivier, good outcrops of red and green coarse sandstones with intercalated finer grained sandstones occur in road cuttings. The sandstones show well developed cross bedding testifying to their fluvial origin. The bridge over the Lewerrivier is crossed 6 km after the first outcrop of Haribes sandstones, and the good outcrops continue after the bridge. Thereafter the landscape becomes quite flat, and only after another 10 km are the sedimentary structures of the Haribes sandstones seen again on the slopes of hills to the N of the road.

Further on, the sediments in road cuttings change colour to more grayish shades, as the base of the Haribes sandstones is approached. The contact to the underlying Zamnarib Member is passed some 22 km after crossing the Lewerrivier, and the sediments are now noticeably finer grained with more mudstone, which points to a more quiet depositional environment. The road then descends into the bed of the Zwillingsrivier, where in some places surficial calcrete can be seen on top of the Zamnarib sediments. The sediments become locally more fine-grained and dark, which goes hand-in hand with a decrease in vegetation. The road continues down the stratigraphy until, after some 18 km from the Zwillingsrivier, it is underlain by the Breckhorn Formation. This formation is composed of red to purple cross-bedded sandstones, which form prominent outcrops. The cross-bedding is formed by rapid deposition of material by heavily laden currents. After another 10 km, the road follows a steep decline into the valley of the Hudup, which marks the contact between the Breckhorn Formation and the Wasserfall Member of the Stockdale Formation. This member consists of a friable cross-bedded red sandstone with minor shale. It is less resistant to weathering than the other sediments and erosion has carved spectacular cliffs out of the Wasserfall sandstone. A good view of the incised river bed into the Wasserfall Member can be seen after crossing the bridge over the Hudup River. After 5 km, the cross bedding of the Wasserfall sandstone can be seen clearly in road cuts, and after another 5 km, the underlying Inachab Member is reached with no noticeable change in the landscape. This member is made up of grey, red and purple sandstones and a little shale.

About 14 km after the Hudup bridge a good view into the Maltahöhe valley is followed by the scenic descent down the Schwarzrand Escarpment. It is marked by the contact between the Inachab Member and the underlying Haseweb Member. The road then follows a valley that has been deeply incised into the friable cross-bedded sandstones and mud-cracked shale of the Haseweb Member. This member also underlies the town of Maltahöhe, which is reached some 113 km from Mariental.

Just outside Maltahöhe, the road continues to descend into the valley of the Kuchab River, and crosses sandstones and shale of the Kabib Member, which underlie the Haseweb Member. They in turn unconformably overlie the rocks of the Schwarzrand Subgroup, which are reached some 2 km outside Maltahöhe. Thereafter the valley opens up and the road takes a turn to the S following the course of the Kaseweb Rivier in an area, which is underlain by Quaternary sediments. The Schwarzrand Escarpment is well exposed to the E with some red dunes accumulated at the base.

Outcrops of greenish shale and sandstone of the Urusis Formation, Schwarzrand Subgroup, can be observed 12 km outside Maltahöhe. The road now follows a fairly wide valley, which is underlain by Schwarzrand sediments. Some red dunes can also be seen. The flat-topped Zaris Mountains representing the Great Escarpment become visible in the far W about 46 km outside Maltahöhe. They are part of the Great Escarpment and are composed of flat lying limestone of the Zaris Formation, Kuibis Subgroup, of the Nama Group overlying basement rocks of the Sinclair Sequence. The landscape remains flat although numerous small rivers cross the road, which dewater eastwards into the Kaseweb River. Further on, the valley narrows and the scenary is dominated in the E by the Schwarzrand Escarpment, with a prominent inselberg on farm Zugspitze, and to the W by the dark limestone of the Zaris Formation. Two prominent isolated hills, composed of Schwarzrand sediments, called the "Schwarzkuppen" resisted the eastward progressing erosion just S of Zugspitze.

As the road continues still underlain by rocks of the Schwarzrand Subgroup, the landscape changes with more undulating hills, although, the Schwarzrand Escarpment remains prominent in the E. On the farm Saraus, just after entering the Bethanien District, spectacular inselbergs of Schwarzrand Subgroup sediments overlain by beds of the Fishriver Subgroup can be seen to the E and the W of the road. Thereafter the road makes a westward turn away from the Schwarzrand Escarpment, crosses the Konkiep River and just before Helmeringhausen, a marked change in the landscape to an increasingly rugged topography is indicative of a change in the geology, and to the W of the road dark lavas of the Barby Formation and shale of the Kunjas Formation of the Mesoproterozoic Sinclair Sequence can be seen. The Sinclair Sequence accumulated in the Helmeringhausen-Solitaire area during three broad cycles of volcanism, plutonism and sedimentation some 1200 to 900 Ma ago. The area surrounding Helmeringhausen is underlain by rocks of the Sinclair Sequence extensively intruded by subvolcanic granites associated with the Sinclair episode. The Helme River is crossed just before

Helmeringhausen, and just after Helmeringhausen, hills showing typical onionskin weathering are composed of the red Rooikam Granite. This granite intruded the rocks of the Sinclair Sequence some 1200 Ma ago. The flat-topped hills some 10 km outside Helmeringhausen are also composed of Sinclair Sequence rocks, and thereafter the road once again follows terrane underlain by the same granite.

On the farm Gamochas, some 20 km outside Helmeringhausen, the gentle and boulder-strewn hills are composed of quartz porphyry and mafic rocks of the Nagatis Formation, Sinclair Sequence. Further on, sediments of the Kuibis Subgroup of the Nama Group again form flat-topped hills to the E of the road, and the pink Tumuab Granites of the Tiras Mountains become visible in the SW. The Tumuab Granite intruded the Sinclair Sequence at roughly the same time as the Rooikam Granite, and weathers to produce large boulders. As one leaves the area underlain by these granites, the valley becomes extremely wide and provides, in the E, a good view of the lower Nama Group unconformably overlying the gneisses and amphibolites of the Namaqualand Complex, which accumulated in a sea adjacent to the Kaapvaal Craton, and underwent high-grade metamorphism some 1200 Ma ago. Across the peneplain, the dry riverbeds are marked by the occurrence of large Camelthorn trees. Some 55 km S of Helmeringhausen, the Namib Sand Sea becomes visible and further to the W; some distant mountains are composed of a variety of rocks of the Namaqualand Complex.

The road now follows the wide peneplain and for the next 47 km the view is dominated by the Namib Desert in the W with, well visible in the E, the Great Escarpment with the Nama Group overlying the Namaqualand Complex. Approaching the main B4 road from Keetmanshoop to Lüderitz, some 5 km from the junction, biotite-rich augen-gneisses and light granite gneisses of the Namaqualand Complex form hills to the W of the road, and just next to the junction an isolated small hill of the same rocks shows well developed onionskin weathering. The same rocks are also exposed in a few road cuttings next to the main road, as one travels the remaining 2 km to Aus.

## 9.9 Mariental – Keetmanshoop – Lüderitz

This 550 km long route starts of in rocks of the Karoo Sequence, then descends down through the Nama Group, passes the Namib Sand Sea and ends in rocks of the Namaqualand Complex (Figs 9.9.1, 9.9.2). It covers more than 1000 Ma of Namibian geological history, but for more than 200 km of the total distance of 224 km from Mariental to Keetmanshoop the road follows an extremely flat peneplain, which is underlain by the Permo-Carboniferous Dwyka Formation of the basal Karoo Sequence.

The Dwyka Formation, comprising tillite and various conglomeratic, fluvio-glacial and glacio-marine deposits with interbedded sandstones and shale averages a thickness of about 200 m. It has been established that glacial deposits in the South Kalahari Basin, where the Keetmanshoop area is situated, were transported by two separate ice sheets, one coming from the N, and the second, slightly younger, coming from Griqualand in

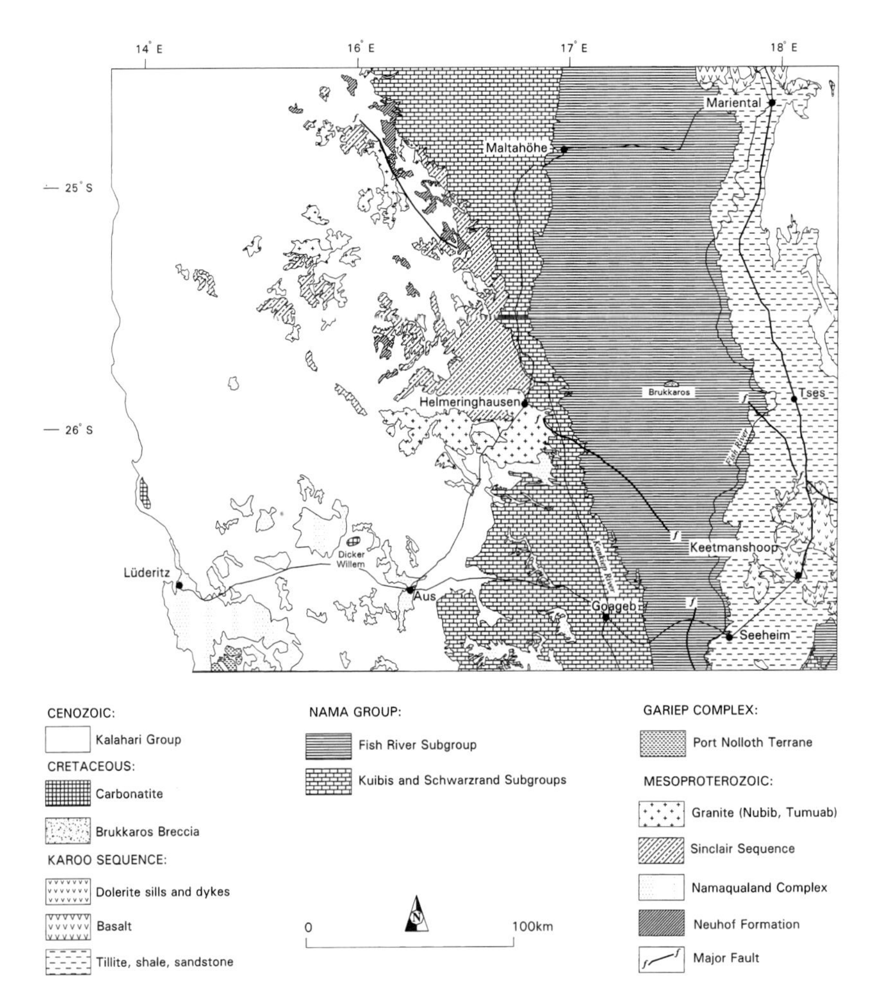


**Fig. 9.9.1:** Geological map for route 9.9.

the E (Du Toit 1921), as evidenced by glacial striations on the underlying Fish River Subgroup sandstone on farms Nanebis 120 and Rheinfels 125 SW of Keetmanshoop. The succession begins with a tillite considered to be a consolidated ground moraine. The matrix is bluish-gray calcareous mudstone. Enclosed in the groundmass are angular rock fragments, many of which are facetted and striated. About 90 % consist of lo-

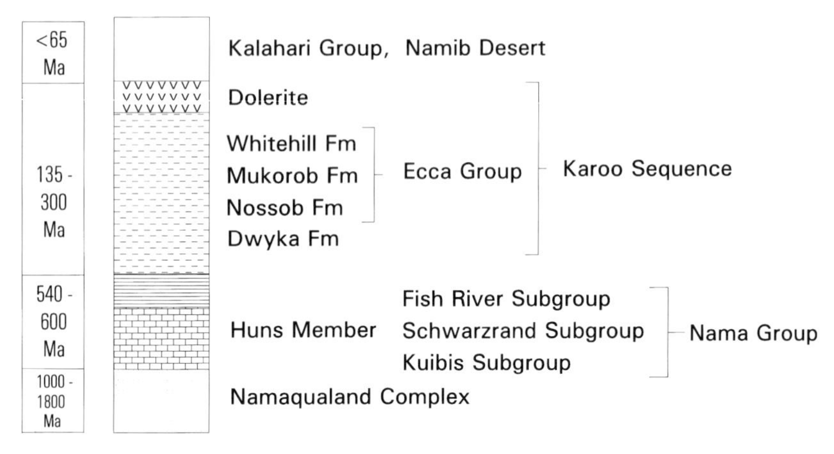


**Fig. 9.9.2:** Stratigraphic column for route 9.9.

cally derived sandstones of the Fish River Subgroup, Nama Group. It is probable that the reddish tint often observed in the matrix is caused by finely disseminated rock flour derived from local material. The upper Dwyka Formation, which by the occurrence of fossils has been recognised to be of marine origin, has been subdivided into seven members and is comprised of shale, siltstones and sandstones (Genis & Schalk 1984).

The prominent Weissrand Plateau appears in the E some 7 km S of Mariental, formed by the Nossob Sandstone and the overlying basal Kalahari calcrete. A turn-off to the former erosion relic of Mukorob (see 8.20) is passed 97 km S of Mariental near Asab.

The Brukkaros Mountain (see 8.2) breaks the monotony and becomes visible 90 km S of Mariental. The road now passes an area renown for the Gibeon Meteorite shower (see 8.11). The area is also known for the occurrence of numerous kimberlite pipes, all of which, however, were found to be not diamondiferous. As one proceeds, the Weissrand Plateau retracts to the E, and becomes completely invisible some 100 km S of Mariental, although it reappears S of Asab.

Near Tses, E of the road, a prominent dune of Kalahari sands can be seen, illustrating past aeolian activity. S of Tses the Weissrand Plateau disappears towards the E, while the Brukkaros Mountain in the W remains visible almost all the way to Keetmanshoop. Some 55 km before Keetmanshoop, the contact between the Dwyka Formation and the overlying Ecca Group is passed, but here this contact is quite indistinct. In the N it is marked by the Nossob sandstone, but this sandstone thins in the Keetmanshoop area and is widely substituted by a layer of shale.

Approaching Keetmanshoop, the first Karoo dolerites, which are so characteristic of the Keetmanshoop area, can be observed. Two large superimposed dolerite sheets intruded the sedimentary successions to the N and NE of Keetmanshoop. The dolerite is usually fine-grained and consists essentially of plagioclase and pyroxene. Druses covered with natrolite crystals have been found in some places. The lower sheet was generally emplaced between sediments of the Dwyka and Ecca successions. The upper sheet extends over an area of about 670 km$^2$ and intruded at the level of the Whitehill Formation of the upper Ecca Group. After intrusion, the dolerite sills were fractured, faulted and subsequently intruded by dolerite dykes. The dolerite has a Jurassic age of about 180 Ma. Until Keetmanshoop is reached, numerous outcrops of Ecca Group sediments and Karoo dolerite occur in road cuttings. The prominent Dolerite Hills (see 8.5) can be observed in the distance to the east of the road.

Towards Keetmanshoop the dolerites form distinct hills in the vicinity of the town, and small outcrops of the Whitehill Formation are typical for the hilly area NE of Keetmanshoop, where these can be observed in road cuttings in contact with the dolerite some 32 km and 17 km outside the town. The Whitehill Formation consists of dark carbonaceous shale which has a conspicuous white colour when weathered, giving the Formation its name. It contains the lizard-like fossil *Mesosaurus* (see 5.). This has provided early proof of the continental drift as proposed by Alfred Wegener and Alex du Toit at the beginning of the 20$^{th}$ Century because of a subsequent find in the Irati Shale of Brazil. Some well-preserved specimens of *Mesosaurus* can be viewed on farm Gariganus 157.

As one leaves Keetmanshoop, the Karas Mountains (see 8.14) can be seen in the distance in the SE. A good example of dolerite weathering occurs about 1 km outside Keetmanshoop next to the turn-off to Grünau. The uppermost part of a dolerite sill has completely disintegrated to form well-rounded dolerite boulders.

The road now follows a peneplain with little or no outcrop, but in the distance flat-topped hills composed of Dwyka Formation sediments and immediately N of the road some dolerite-capped hills containing Ecca Group sediments at their base can be seen. Some 6 km outside Keetmanshoop road cuttings display dark Dwyka Formation sediments for about 20 km, and after 26 km, a table-topped mountain composed of sediments of the Fish River Subgroup of the Nama Group is passed to the S, while to the N there is a prominent row of dolerite-topped mountains.

The road again follows a peneplain for the next 15 km, still underlain by Dwyka Formation sediments. Just after the turn-off to the Fish River Canyon (see 8.9) at Seeheim, the road starts to descend rapidly into the valley of the Fish River cutting into the upper Fishriver Subgroup sandstone and shale of the Nama Group, which underlies the Dwyka Group. The Nama Group covers large parts of southern Namibia and was deposited in a shallow marine environment during the late stages of the Damaran orogeny as a molasse derived from the erosion of the uplifted Damara and Gariep mountain belts.

The Fish River is one of the major ephemeral rivers of Namibia, which in certain parts carries water almost all year round, and is crossed some 41 km from Keetmanshoop. The road climbs out of the Fish River valley for the next 8 km, and the purple, pink and whitish sandstone and shale of the upper Fish River Subgroup are well exposed in road cuts. The fairly high table mountains seen on either side of the road also comprise of Nama Group sediments. The road then proceeds in successively lower stratigraphic levels of the Nama Group, thereby providing a complete profile through this Neoproterozoic to Early Cambrian Sequence.

Some 14 km after the Fish River crossing, horizontally layered, pink upper Fish River Subgroup shale are found in road cuts. The road then crosses a peneplain, underlain by reddish sandstones of the upper Fish River Subgroup, and flat-topped mountains composed of pink shale can be seen on the northern side of the road. After another 12 km the road descends into the red sandstone and shale of the middle Fish River Subgroup. It then follows a peneplain underlain by these rocks, and progresses, after another 11 km onto red shale and sandstone of the lower Fish River Subgroup. The table-mountains in the vicinity are composed of the same shale and sandstone. During the descent into the valley of the Guib River, black limestone occurs in road cuttings on both sides of the road some 4 km further on, and marks the approach to the Schwarzrand Subgroup.

After crossing the Guib River, black limestone cliffs of the Huns Member of the Schwarzrand Subgroup become prominent on the southern side of the road. The road then remains in limestone, quartzite and shale of the Schwarzrand Subgroup, until it crosses the Konkiep River at Goageb some 18 km after the Guib River crossing. The table mountains on either side of the road are composed of Schwarzrand Subgroup sediments, while the wide valley of the Konkiep River, a major tributary of the Fish River, is filled with Cenozoic fluviatile sediments.

For the next 20 km the road crosses a vast peneplain underlain by shale, quartzite and limestone of the basal Kuibis Subgroup of the Nama Group, until some 22 km after the Konkiep River, the plateau of the Schwarzkuppe, capped by black limestone overlying gneiss of the Namaqualand Complex appears in the N. This gneiss formed from sediments that accumulated in a sea adjoining the Kaapvaal Craton, which underwent high-grade metamorphism some 1200 Ma ago. To the W of the Schwarzkuppe plateau and N of the road, the canyon of the Bree River is incised deeply into sediments of the Kuibis Subgroup and the gneiss of the Namaqualand Complex, which forms the basement to the Nama Group.

The road then follows for 30 km a peneplain underlain by black limestone and pink shale of the Kuibis Subgroup, which are exposed in occasional road cuts. After a slight ascent within the Kuibis Subgroup sediments, the road finally descends into a valley underlain by Cenozoic alluvium some 45 km before Aus. The valley is flanked by table-topped mountains composed of Kuibis Subgroup sediments resting unconformably on the gneisses of the Namaqualand Complex. Cenozoic sands of the Namib Desert can be seen accumulating on the western slopes of the mountains, particularly to the S of the road.

Two prominent hills of gneisses of the Namaqualand Complex occur about 20 km outside Aus, just N of the road, while to the S, the Huib Hochplateau composed of the Kuibis Subgroup underlain by the rocks of the Namaqualand Complex represents the

Great Escarpment in the area. It is in this area, that some of the oldest fossils known on Earth have been found (see 5.). An isolated small hill, 3 km outside Aus, after the turn-off of road C13 to Helmeringhausen, is composed of light granitic gneiss of the Namaqualand Complex and shows well developed onionskin weathering. The same rocks are also exposed in a few road cuttings next to the road before Aus, which is reached some 208 km after Keetmanshoop.

The 122 km long road from Aus to Lüderitz follows the railway line, where the first diamond was discovered in 1908, and directly opposite the turn-off to Aus, the granitic gneisses of the Namaqualand Complex are cross-cut by pegmatitic dykes. Further on, these rocks display typical onionskin weathering. After Aus, the road enters a peneplain underlain by Cenozoic sediment, and the mountains to the S of still represent the granitic gneisses of the Namaqualand Complex.

After some prominent sand dunes of the Namib Sand Sea, the eastern boundary of the former so-called “Sperrgebiet” or Diamond Area No 1 is passed some 15 km from Aus. The “Sperrgebiet” was declared a safety zone by the German Imperial Government in 1909 to protect Namibia’s most valuable raw material, the diamond. This area today remains controlled and can only be entered with a permit. The dunes in the area are fairly stable, partly vegetated, and some of the larger trees present indicate limited underground water.

The prominent Dicker Willem carbonatite mountain appears some 24 km outside Aus and belongs to the southern line of anorogenic intrusive rocks extending from the coast S of Lüderitz in a northeasterly direction towards Brukkaros, and has an age of approximately 135 Ma. Smaller hills of marble and amphibolite of the Namaqualand Complex are seen in the S, and this complex dominates the basement geology all the way to Lüderitz. The road then follows the Namib Desert peneplain, which is only broken by some hills composed of rocks of the Namaqualand Complex, which can be seen in the distance. The first noticeable hill occurs after 35 km from Dicker Willem at the Tsankaib Siding, and is composed of layered gneisses of the Namaqualand Complex. The layered structure is well revealed by the sand accumulated on its flanks, and the mountains in the southern distance also belong to the Namaqualand Complex.

The road takes a marked turn to the southwest some 10 km after Tsankaib, and to the NW a nice view into the desert is revealed. A large field of parabolic dunes becomes visible some 22 km after Tsankaib, and is reached after another 7 km. To the south of the road good examples of Namib dunes burying gneiss of the Namaqualand Complex can be seen, and frequently, these sands encroach onto the road, which explains the occasional occurrence of road signs with the warning “Sand!” (Fig. 9.9.3). Large parabolic dunes of the Namib Sand Sea dominate the landscape for the next 25 km, before some larger mountains composed of amphibolite of the Namaqualand Complex become visible in the NW. The ghost town of Kolmanskop, some 10 km outside Lüderitz, is a remainder of the haydays of the diamond rush (see 8.15). A museum, guided tours and a walk through the town provide an opportunity to understand the environment, where the first diamond was found.

**Fig. 9.9.3:** Roadsign on the road to Lüderitz.

Between Kolmanskop and Lüderitz there are ample outcrops of amphibolite with cross-cutting leucocratic dykes of the Namaqualand Complex next to the road. The town itself is underlain by granitic gneiss of the Namaqualand Complex, which form the surrounding hills, such as Diamantenberg on which the church is situated. The peninsula sheltering the Lüderitz harbour is composed entirely of gneiss of the Namaqualand Complex, and scenic drives to Agate Beach or Diaz Point all remain in the area underlain by these rocks.

## 9.10 Aus – Rosh Pinah – Oranjemund

by Volker Petzel and Gabi Schneider

This southern Namibian route follows the eastern boundary of the former "Sperrgebiet" or Diamond Area No 1. The "Sperrgebiet" was declared a safety zone by the German Imperial Government in 1909, to protect the diamonds discovered along the coast and to this day entry is strictly controlled and can only be done with a permit. The mining town of Rosh Pinah is reached after some 170 km, from where a 100 km long road leads to the Namibian diamond capital of Oranjemund (Figs 9.10.1, 9.10.2). A special permit is required to travel to Oranjemund, which must be obtained from Namdeb Security and the Protective Resources Unit of the Namibian Police. Please note that these permits have to be applied for well in advance of the intended travel date! The Orange River, just S of Rosh Pinah, can be visited by taking a 20 km long detour to Sendelingsdrif. The river can, however, not be crossed, as there is no borderpost with South Africa.

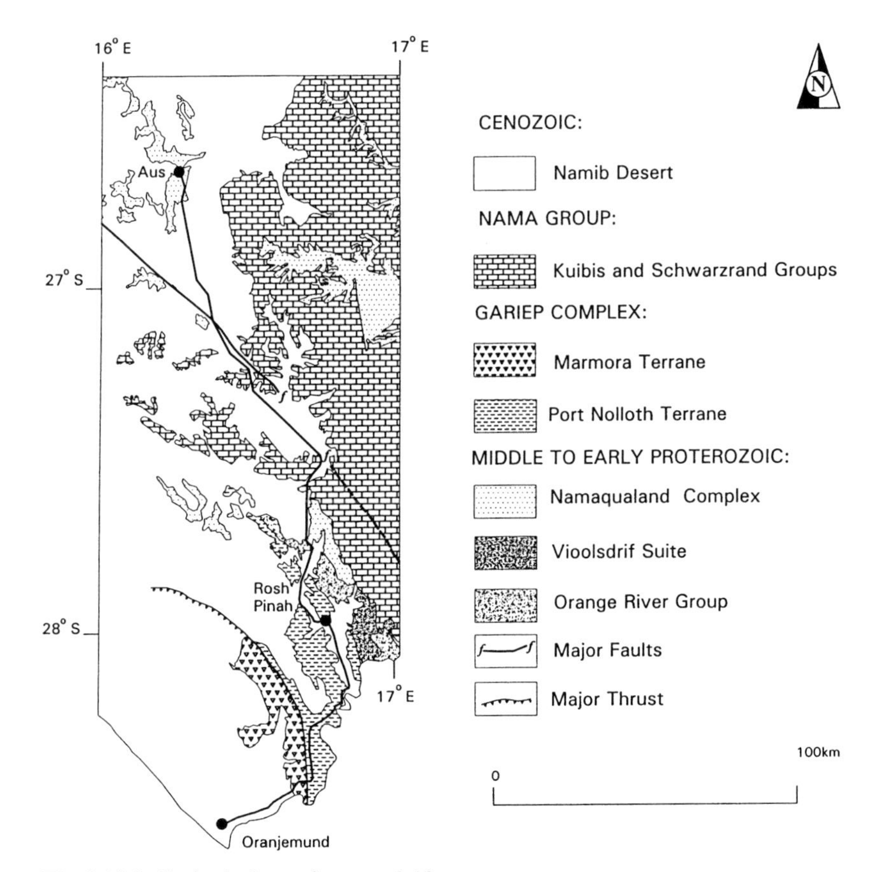


**Fig. 9.10.1:** Geological map for route 9.10.

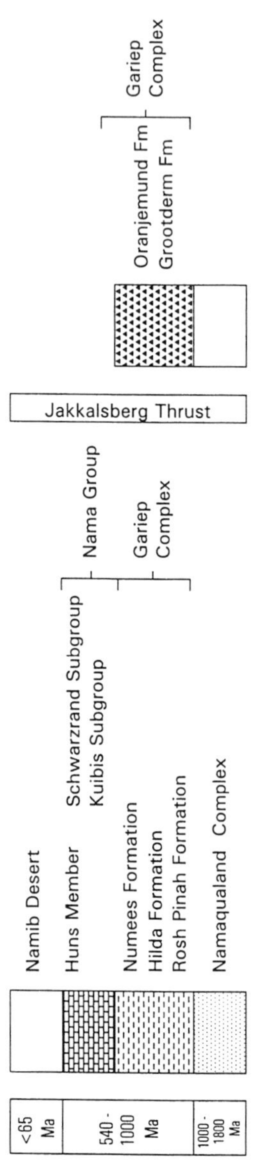


**Fig. 9.10.2:** Stratigraphic column for route 9.10.

The route leaves Aus straight to the S in an area that is underlain by granitic gneisses and biotite-rich augen gneisses of the Namaqualand Complex. After 5 km some outcrops of Namaqualand Complex granitic gneiss occur very close to the road. The granitic gneiss is very inhomogenous, typical of gneiss formed by partial melting, with some outcrops displaying the typical onionskin weathering, whilst elsewhere it is more jointed. Once these outcrops are passed the road follows a peneplain underlain by Cenozoic sediments. The rocks of the Namaqualand Complex accumulated in a sea adjoining the Kaapvaal Craton, and underwent high-grade metamorphism some 1200 Ma ago. Smaller hills comprising gneiss and schist of the Namaqualand Complex appear in the distant W, to the east the Great Escarpment represented here by the prominent Huib Hochplateau of the basal Kuibis Subgroup, Nama Group, overlying gneisses of the Namaqualand Complex, dominates. It is in this area that some of the oldest fossils known on Earth were found in sediments of the Nama Group (see 5.). Rocks of the Nama Group were deposited in a shallow marine environment during the late stages of the Damaran Orogeny as a molasse derived from the erosion of the uplifted Damara and Gariep mountain belts.

The appearance of the plateau in the E changes some 25 km outside Aus. While previously light coloured sandstones formed the top layer, it now is capped by a layer of dark limestone, which is clearly visible. The disconformity between the Nama Group sediments and the underlying gneisses of the Namaqualand Complex is also well exposed, and smaller hills, comprising gneisses and schists of the Namaqualand Complex, are still visible in the distant W. Some 48 km from Aus, the dark Vaalkop, composed of biotite gneiss and granodioritic augen-gneiss of the Namaqualand Complex appears prominent in the W. After another 10 km, the road passes a major fault zone in the E, along which strata of the Namaqualand Complex and the basal units of the Nama Group are downfaulted. Consequently, the flanks of the plateau which have so far displayed the contact between the Namaqualand Complex and the Nama Group, from now on only show Nama Group strata. The road here passes through a small mountain range comprising the last outcrops of gneisses and amphibolites of the Namaqualand Complex at their base, overlain by Nama Group sediments. Thereafter the road moves out onto a peneplain underlain by Cenozoic sediments.

The Huib Hochplateau is still prominent in the E some 62 km from Aus, but it is now composed of quartzites, shales and siltstone of the Schwarzrand Subgroup, Nama Group, the subgroup overlying the Kuibis Subgroup. In the W are the Swartkloofberge comprising limestones of the the Schwarzrand Subgroup and shales, quartzites and dark limestones of the Kuibis Subgroup. The sediments in the E and in the W are separated by a major fault along which the western block has been uplifted. There is intermittant production of marble from two quarries in the area, and the turn-off to the Swartkloofberg Marble Quarry is passed 84 km S of Aus. In the E, the Huib Hochplateau now displays a lighter colour on top, caused by a capping of yellowish limestone. The road now follows a very wide valley underlain by Cenozoic sediments, and the Nama Group sediments are only visible in the distance.

The road makes a marked turn to the SW some 108 km south of Aus, and here limestones of the Schwarzrand Subgroup occur on the western side of the road. After 10 km the road makes another turn to the S and Schwarzrand Subgroup limestones crop out on either side of the road. There are two springs in this area, which are due to the contact of water-carrying sandstones of the Schwarzrand Subgroup with impermeable Schwarzrand limestones. Four kilometers from here, road D 463 turns off to the E in a valley representing a topographic expression of the large fault mentioned above. The road crosses the fault, and now in the uplifted block, the gneisses of the Namaqualand Complex, separated by an unconformity from the overlying Kuibis Subgroup sediments, is once again visible. The Witputzberge in the W are composed of gneisses of the Namaqualand Complex overlain by Neoproterozoic mixtites of the Numees Formation of the Gariep Complex. The Gariep Complex represents the equivalent of the Damara Orogen in southern Namibia. It was deposited in an ocean to the W of the Kaapvaal Craton from 900 Ma ago onwards, and underwent metamorphism during the Pan African event some 600 to 500 Ma ago. In this area, the Gariep Complex comprises the Numees, Rosh Pinah and Hilda Formations.

Some 128 km from Aus the mountains on the eastern side of the road are formed by granitic gneisses of the Namaqualand Complex, whilst in the far E the Nama Group sediments of the Huib Hochplateau are only visible in the distance. After a further 7 km, a high, dark mountain range appears in the SE and is composed of mixtite, dolomite and shale of the Numees Formation of the Gariep Complex overlying the Namaqualand Complex. The road then approaches a ridge of massive gneiss of the Namaqualand Complex, which is crossed after another 5 km providing good outcrops of gneiss cross-cut by pegmatitic dykes. About 148 km S of Aus, the mountain range to the S displays a marked change in lithology towards the SW. The range is composed of gneiss of the Namaqualand Complex in the NE, while in the SW these rocks are overlain by the basal mixtites of the Numees Formation and the felsites of the Rosh Pinah Formation. The felsites have a distinct light colour and the contact is clearly visible in the mountain range.

The road then passes through hills composed of rocks of the Rosh Pinah Formation after another 4 km, and outcrops occur fairly close on either side of the road. After 2 km the road takes another marked turn to the W, and the mountains seen in the distance in the desert also represent rocks of the Gariep Complex. Three kilometers further on, light coloured felsite of the Rosh Pinah Formation appears prominent between darker schists of the same formation on the eastern side of the road. Thereafter the road follows a major fault. To the east the hills are composed of Rosh Pinah Formation, whilst to the W, the mixtites of the Numees Formation form very distinct, high hills displaying sedimentary layering. The hills of the Hilda Formation carbonates can also be seen in the distance in the SW.

Some 4 km before the Rosh Pinah Mine, the valley widens considerably, and is again underlain by Cenozoic sediments. The mountains in the distant W are composed of limestones of the Hilda Formation. The Rosh Pinah Mine, some 170 km from Aus, is

Namibia's main zinc producer, and the entire infrastructure of the Rosh Pinah settlement has been established by the mine. Production started in 1963, and some 500 000 t of ore containing an average grade of 7 % zinc, 2 % lead, 0,1 % copper and some silver are produced annually. The ore is of the sedex type and comprises sphalerite, galena, tennantite, tetrahedrite, chalcopyrite, bornite and stromeyerite. Rosh Pinah is one of the largest zinc mines in Southern Africa, and the mine together with the development of the nearby Scorpion Mine due W of Rosh Pinah, as well as the zinc mines of northwestern South Africa fall within one of the world's major zinc provinces.

Driving S from Rosh Pinah the road passes the slimes dam of the Rosh Pinah Zinc Mine, which is located just outside the town. After 9 km, at a water pumping substation, the road branches towards Oranjemund, but free access ends after the first 6 km at the security gate of the Diamond Area. Once inside the Diamond Area, the road enters and is bordered by impressive mountain ranges belonging to the Gariep Complex on both sides. To the E, they consist predominantly of diamictites of the glacial Numees Formation, while to the W metasedimentary units of the underlying Hilda Formation are exposed. Between these ranges the road to Oranjemund follows a broad valley underlain by calcrete, sheetwash gravels and sand.

The Orange River is reached some 10 km from the security gate. The river, in the course of its geological and geomorphological history, has played an important role in the development of the unique diamond deposits of the area. Apart from acting as a primary conduit for the diamonds, found within the river gravel itself and along the coast, both onshore and offshore, it also transported downstream from inland Africa enormous sand masses which now contribute to the Namib Sand Sea.

The origin of the Orange River can be traced back to the breakup of Gondwanaland, when, in a humid period during the Cretaceous, erosion was at its peak in central southern Africa. It has been estimated by Hawthorne (1975) that during this period 1400 m of sedimentary rocks were eroded and transported by rivers into the Atlantic Ocean. Kimberlites, which are ubiquitous in central southern Africa, were eroded preferentially during their crater phase, and the diamond-bearing detritus was transported downstream by the Karoo and Kalahari Rivers, draining most of the interior of southern Africa at this time (De Wit 1999). The Kalahari River, the predecessor of the Orange, was a slow flowing, meandering river that dumped millions of tons of mud and plant material into the sea. It was during this phase that the meandering character of the present-day Lower Orange was established, but, the low energy was not sufficient to transport clasts and diamonds far from their source area.

Tectonic uplift, about 60 million years ago, resulted in the capture of the upper Karoo River by the Kalahari River, and the Orange River was established in more or less its present-day configuration (De Wit 1999). Drier periods followed with a dramatic reduction in erosion and a concomitant decrease in sediments carried to the sea, which is evident from the Oligocene sedimentary hiatus off the coast. Wetter conditions returned during the Miocene, and erosion increased. Diamonds, eroded from secondary placers and exposed kimberlites were transported seawards by fast-flowing rivers, together with considerable amounts of sand and gravel. The diamonds, sand and gravels were redeposited in sedimentary trapsites, within this river system, and, upon reaching the sea, distributed along the South Atlantic coast by north-flowing longshore currents. As the energy of the longshore currents decreased, they were finally laid down along linear shorelines and in pocket beaches. The action of southerly winds then further upgraded these deposits by locally forming particularly rich deflation and aeolian diamond placers.

As one reaches the Orange River the first mine workings that can be seen are S of the river, within the Republic of South Africa, on the slopes of the Jakkalsberg. The Jakkalsberg Mine processes gravels belonging to both the oldest river terraces of Middle Miocene age (Proto-Orange stage), and the younger 5–2 Ma old terraces of the Meso-Orange. Both terraces contain diamonds, with the younger Meso-Orange being lower in grade, but on average producing larger stones than the Proto-Orange terraces.

The first mine on the northern Namibian side of the Orange River is the Sendelingsdrif Diamond Deposit, which predominantly consists of Proto-Orange terraces. Here, the road turns to the W, following the northern river bank, and after 6 km it crosses the Proto-Orange terrace of the Obib Diamond Deposit which, like the Jakkalsberg Deposit, consists of both Meso- and Proto-Orange terraces. The former is located closer to the river, and at a lower elevation. To the N of this vantage point diamictites, greywackes, phyllites and minor limestones of the Wallekral Formation of the Hilda Sequence are exposed, forming a range of hills.

The road after a further 10 km passes the Daberas Diamond Deposit, where once again both Proto- and Meso-Orange terraces are being exploited. To the N, a thin sliver of Numees Formation diamictite outcrops on top of the Wallekraal Formation which in turn is overlain by greywacke, quartzite, arkose, limestone and dolomite of the Holgat Sequence. A further 15 km downriver the turn-off to the Auchas Diamond Deposit is reached, where diamonds are mined from Proto-Orange terraces.

The Jakkalsberg Thrust, a major tectonic feature, which marks the transition between the Port Nolloth and the Marmora Terranes of the Gariep Belt, is traversed some 9 km from the Auchas turn-off. Along this thrust the two terranes, distinguished both by tectonic style as well as lithology, are in juxtaposition, with the Marmora Terrane being overthrust onto the Porth Nolloth Terrane towards the east. In this area, the former consists predominantly of chloritic schist, amphibolites and ultramafic rocks, while further to the W dolomites of the Grootderm and Oranjemund Formations predominate. Dolomites of this tectonostratigraphic unit also form the westernmost outcrops seen along the road to Oranjemund.

Eight kilometres after crossing the Jakkalsberg Thrust another road turns off to the Arriesdrif Deposit, where Proto-Orange terraces are mined for diamonds. Moreover, the Arriesdrif and Auchas localities are famous for their rich Cenozoic fossil finds (see 5.), which include several species of *Prohyrax*, the forerunner of the present-day *Hyrax* (Pickford 1994), as well as the fossil crocodile *Crocodylus gariepensis* (Pickford 1996), teeth of the primitive elephant *Gomphotheres*, and a complete skull of *Eozy-*

*godon morotoensis*. A human skull, which has become known as the "Orange River Man" was also found by Pickford, who described it as an archaic *Homo sapiens* and dated it at 50 000 to 100 000 years. The skull is on display in the museum of the Geological Survey of Namibia in Windhoek. Leaving Arriesdrif behind, the road, which is underlain by sand and sheetwash deposits, again approaches the Orange River and follows it to Oranjemund.

The history of diamond mining and the establishment of Oranjemund dates back as early as 1884, when german geologists sunk a shaft beside the Orange River to examine the sedimentary layers for gold and diamonds. In 1910 further pits were excavated in the area under the supervision of geologist Dr. Ernst Reuning, but as before, this exploration phase was again unsuccessful. It was not until 1928, that another attempt was made by Consolidated Diamond Mines (CDM) geologists, who returned to the area after Dr. Hans Merensky, together with Reuning, had discovered rich diamond placers to the S, at Alexander Bay in South Africa. This time the prospectors were successful and diamond-rich gravels were found, surprisingly less than 100 m from the barren pits dug in 1910. This discovery spurred further exploration, which was followed by intensive mining operations, especially as the diamond deposits at Oranjemund were found to represent the world's largest diamond placers. They have the highest ratio of gem quality to industrial diamonds, with 98 % of the recovered stones being of gem quality. The town of Oranjemund was established inland of the Orange River mouth, and the head office of CDM moved from Lüderitz to Oranjemund in 1943. In 1994 a new company, Namdeb Diamond Corporation, was formed, which is owned in equal parts by De Beers Centenary and the Namibian Government.

## 9.11 Seeheim – Ai-Ais

The route begins at the turn-off of road 28 from the main Keetmanshoop – Lüderitz road, just before the Fish River is crossed; and leads to the Fish River lookout point and the hot springs at Ai-Ais. It is 171 km long and starts off in an area underlain by the Karoo Sequence, then crosses rocks of the Nama Group to end in Mesoproterozoic Namaqualand Complex terrane (Figs 9.11.1, 9.11.2).

Initially, the terrane is underlain by sediments of the Dwyka Formation and shows limited relief. It is only broken in the E by some hills composed of Karoo dolerite and the Klein Karasberge in the distance, and the valley of the Fish River barely recognizable in the W. The wide floodplain of the Löwen River, which dewaters into the Fish River, is crossed after 26 km. Due E of the road the Löwen River has been dammed to form the large Naute Dam, the major source of water supply in that area. Thereafter, the road continues in an area underlain by the Dwyka Formation. However, after some 40 km, there is a marked change in landscape, and hills composed of shale and limestone of the Kuibis and Schwarzrand Subgroups of the Nama Group appear immediately E of the

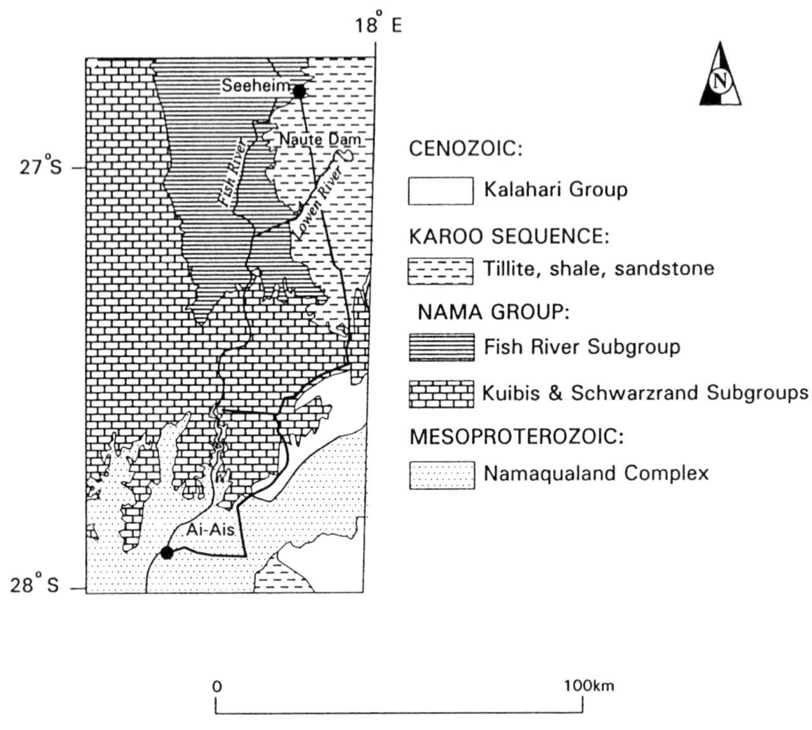


**Fig. 9.11.1:** Geological map for route 9.11.

**Fig. 9.11.2:** Stratigraphic column for route 9.11.

road. They are deeply transsected by seasonal rivers dewatering westwards into the Fish River. An example is the Holoog River.

Road 601 turns off 71 km S of Seeheim in a southwesterly direction. To the SE, the Holoog Berge, consisting of gneiss of the Namaqualand Complex and table-topped mountains comprising sandstone, shale and limestone of the Kuibis Subgroup, dominate the view. The peneplain in the NW contains the deepening Fish River Canyon, which, although not seen, can only be imagined. After 30 km, there is a turn-off to the W, which leads across the Huns Plateau, underlain by Kuibis Subgroup rocks, to the Fish River Canyon lookout point (see 8.9).

After the turn off, the road continues along the western foot of the Kuibis Subgroup mountain range, before it enters terrane underlain by the Mesoproterozoic Namaqualand Complex after crossing the Rosyntjiesbos River. Gneiss of the Namaqualand Complex forms the Kuduberg Range to the E of the road, while to the W, the sandstone and limestone of the Kuibis Subgroup underlie the slope towards the Fish River Canyon.

About 20 km from the turn-off to the lookout point, mountains occur closer to the road in the E and the W. They are composed of pink gneisses, amphibolites and biotite schists, intruded by megacrystic granites, all belonging to the Namaqualand Complex. Prominently visible in the E are the Kanebis Mountains, composed of granite, and in the W the Hochstein, composed of amphibolite.

The road joins the C10 road from Grünau 43 km after the lookout point turn-off and now runs in a westerly direction remaining for 12 km in terrane underlain by gneiss. For the last 9 km to Ai-Ais it crosses a rugged area, where meta-gabbro and a late tectonic granite have intruded the gneisses. Lenses of rose quartz are common and can be observed next to the road.

Ai-Ais lies in the valley of the Fish River, which here has incised its bed deeply into the rocks of the Namaqualand Complex. The water of the hot spring at Ai-Ais is derived from deep-seated fractures within these rocks, formed during the course of extensional tectonics during the break-up of Gondwanaland. These fractures, as well as older ones have also guided the incision of the Fish River. The spring water has a temperature of 60 °C, a pH of 8,75, contains 2030 ppm total dissolved solids, and is of therapeutic value.

## 9.12 Keetmanshoop – Grünau – Noordoewer

This 304 km long route transgresses from Karoo Sequence rocks into the older sediments of the Nama Group, before it follows terrane underlain by the Mesoproterozoic Namaqualand Complex. S of Grünau Karoo Sequence rocks once again prevail, but closer to Noordoewer the Palaeoproterozoic rocks of the Vioolsdrif Suite and Orange River Group appear prominent in the E (Figs. 9.12.1, 9.12.2)

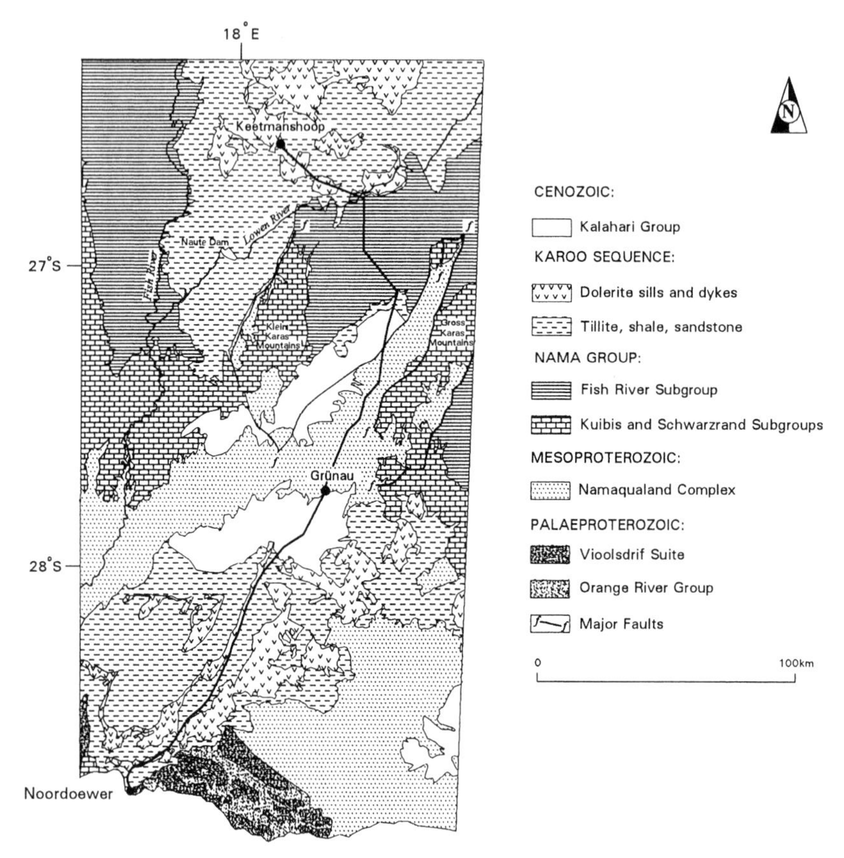


**Fig. 9.12.1:** Geological map for route 9.12.

The road to Grünau leaves Keetmanshoop in a southeasterly direction in an area, which is underlain by sediments of the Dwyka Formation of the Karoo Sequence, which were deposited under glacial conditions some 300 Ma ago. The Dwyka rocks are partly capped by Karoo dolerites, which form the typical table-topped hills seen to the N of the road. One isolated hill to the E is shaped like a pyramid, again capped by dolerite. Meanwhile, the Karas Mountains (see 8.14) are prominently visible towards the S and SE, and some 20 km S of Keetmanshoop the road crosses the Huns River. This river crossing is followed after another 17 km by the Guruchab River, and after another 5 km

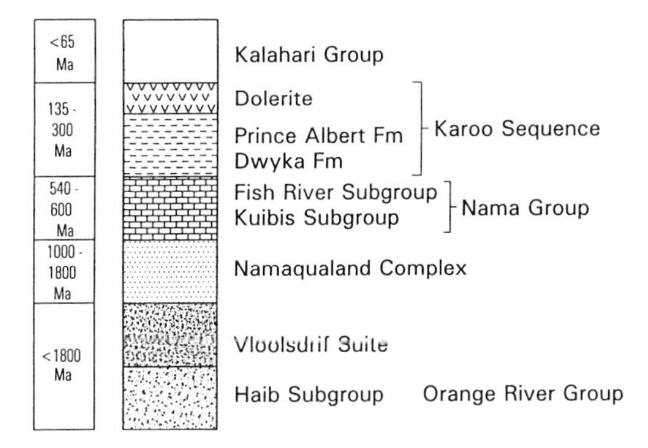


**Fig. 9.12.2:** Stratigraphic column for route 9.12.

the Löwen River. All three rivers are large ephemeral rivers dewatering into the Fish River. The Löwen River is dammed downstream to form the large Naute Dam, the major source of water supply in that area.

Further on, there is a considerable change in the landscape, with a marked increase in relief, as the road follows a valley with the Klein Karas Mountains to the W and the Gross Karas Mountains to the E. The Karas Mountains were formed by reverse faulting under compressional conditions some 500 Ma ago, and more recently in post-Karoo times. Rocks of the Fish River Subgroup of the Nama Group underlie the area for the next 32 kilometers, and can be observed in outcrops along the road. The road then climbs to higher stratigraphic levels, and passes outcrops of the Dwyka Formation and shales of the Prince Albert Formation of the Karoo Sequence. The shales form a monotonous, khaki-coloured succession, with gray limestone lenses.

As the road crosses a peneplain underlain by Cenozoic sediments, it approaches the Gross Karas Mountains, and the highest peak, the Schroffenstein, dominates the view to the E with an elevation of 2202 m above sea level. The Schroffenstein is composed of highly sheared amphibolites and amphibole schist of the Namaqualand Complex. The road crosses on farm Dassiefontein a major block fault, along which rocks of the Namaqualand Complex and the Nama Group were upthrown to form the impressive range of the Gross Karas Mountains. On the eastern side of this fault, the terrane is underlain by coarse-grained granite gneiss, biotite-garnet-sillimanite gneiss and biotite-garnet gneiss of the Namaqualand Complex. The coarse-grained granite gneiss forms the hills on both sides of the road, while gneiss and schist underlie the lower lying areas. The crest of the Karas Mountains seen in the E is formed by conglomerates of the Kuibis Subgroup of the Nama Group.

Some 25 km after Dassiefontein, a prominent ridge on the western side of the road is composed of biotite-garnet-sillimanite gneiss of the Namaqualand Complex. The

Rooiberg and Witberg, some 10 km further on, are again made up of coarse-grained granite gneiss. At the boundary between the Keetmanshoop and Karasburg Districts, the view to the SW is particularly scenic, whilst to the N, the view is dominated by the Karas Mountains. As the southernmost peak of the Karas Mountains is passed, the road enters an area with little morphology, which is underlain by a zone of mixed gneisses of the Namaqualand Complex. Pegmatites have intruded this zone, and are mined for rose quartz on farm Mickberg.

Grünau is reached 160 km S of Keetmanshoop and is underlain by extensive calcrete plains. S of the village, the road follows a peneplain with little or no outcrop, which is underlain by Cenozoic sediments. The only morphological features seen for the next 30 km are linear dunes on the eastern side of the road. Thereafter, the road approaches terrane underlain by shale and mudstone of the Dwyka and Prince Albert Formations, Karoo Sequence. These have been intruded extensively by Karoo dolerites, which form hills on both sides of the road. Two impressive mountains occur on either side of the road some 47 km S of Grünau. The Norachaskop in the W and the Ysterkop in the E have an elevation of some 1130 m above sea level, and consist of Karoo sediments capped by Karoo dolerite. The road continues southwards following a ridge in the E, which is also composed of dolerite.

This dolerite is mined for agate on the farm Ysterputz, about 15 km S of Norachaskop and Ysterkop. It is a rare, banded, light blue, grayish blue and white agate, which occurs as vein fillings in almost vertical, well developed joints in the dolerite. Individual agate bands average 2 to 3 cm in width, but are restricted to a 50 cm wide zone. The agate is marketed as "blue lace agate".

S of Ysterputz the road descends into the valley of the Haib River, a tributary of the Orange River. This valley is still underlain by shale and mudstone of the Prince Albert Formation, intruded by dolerite sills, and dolerite also forms the ridges seen on both sides of the road. However, about 40 km S of Ysterputz, the mountains to the SE have a different morphology, as they now consist of granite of the Vioolsdrif Suite and dacite and rhyodacite of the Haib Subgroup of the Orange River Group. These Palaeoproterozoic rocks have an age of 2000 to 1760 Ma and host the Haib copper deposit, a huge but low-grade porphyry copper deposit, which contains in excess of a billion tons of ore. This might at some stage be developed into one of the largest open pit mining operations on the African continent. Leaving the highly dissected dolerite hills of Tandjieskoppe in the NW, the road descends rapidly into the valley of the Orange River in terrane that is still underlain by Karoo sediments and dolerite.

The Orange River is one of the very few perennial rivers in Namibia and flows from the Lesotho highlands in eastern southern Africa to the Atlantic Ocean at Oranjemund, a distance of some 2 200 km. The large catchment area covers 50 % of South Africa, 95 % of Lesotho, and 15 % of the surface area of Namibia. The Orange River drainage developed in the Cretaceous, following the break-up of Gondwanaland and the associated continental uplift. At that time, tropical conditions prevailed, and the Orange was a huge, meandering river depositing mud and sand into the Atlantic Ocean. Trapped organic material in the offshore sediments through time produced the Kudu Gas Field, which was discovered in 1973, and is now being developed.

At about the same time, diamond-bearing kimberlite pipes were emplaced in central southern Africa. However, their emplacement was followed by rapid erosion under the tropical climate, which liberated the precious stones. About 60 Ma ago, widespread continental uplift of southern Africa caused the Orange River to deeply incise its bed into the old African land surface. This changed the Orange from a mud-carrying, slow moving river to a gravel- and sand-charged torrent. This high-energy regime was also able to transport the heavy diamonds all the way to the Atlantic coast (see 7.2). When the climate changed to drier conditions, the Orange River, because of its catchment area, remained a tropical linear oasis with its associated flora and fauna (see 5.). The so-called proto-Orange terraces were deposited during this time, some 20 Ma ago. A rise in sea level, caused by the melting of the Antarctic ice sheet some 15 Ma ago, choked the river and thereby preserved the proto-Orange deposits. Sea level fluctuations were abundant during the late Cenozoic, and the Orange River mouth moved back and forth across the continental shelf. Sand transported by the river was further moved northwards up the coast by a long-shore drift and re-deposited on the beaches, from where it was then blown inland, forming the Namib Sand Sea, and also the diamond deposits in beach terraces. Today's Orange River is extensively utilized for irrigation purposes, however, its almost 100 Ma old role in the formation of Namibia's most important mineral deposits, the diamond deposits, makes it Namibia's most precious river (Ward & Jacob 1998). Noordoewer is reached 304 km S of Keetmanshoop.

## 9.13 Grünau – Karasburg – Ariamsvlei

The 161 km long route from Grünau to Ariamsvlei and the additional 16 km to the border with South Africa initially traverses terrane underlain by Permo-Carboniferous to Jurassic Karoo Sequence lithologies until after Karasburg, where it continues in areas underlain by rocks of the Neoproterozoic to Early Cambrian Nama Group (9.13.1, 9.13.2).

Grünau itself is underlain by extensive Cenozoic calcrete terraces, and as the road leaves Grünau, the Gross Karas Mountains (see 8.14) are prominent in the NE. These mountains were formed by compression and faulting some 500 Ma ago, and again more recently in post-Karoo times. The southernmost peaks of the impressive mountain range visible from the Grünau – Karasburg road are composed of biotite-garnet-sillimanite gneiss of the Mesoproterozoic Namaqualand Complex, overlain by conglomerate and shale of the basal Kuibis Subgroup and shale and limestone of the Schwarzrand Subgroup, both of the Nama Group.

Some 14 km out of Grünau, the road enters terrane underlain by shale of the Dwyka Formation, Karoo Sequence, but the landscape is rather flat and there is little or no out-

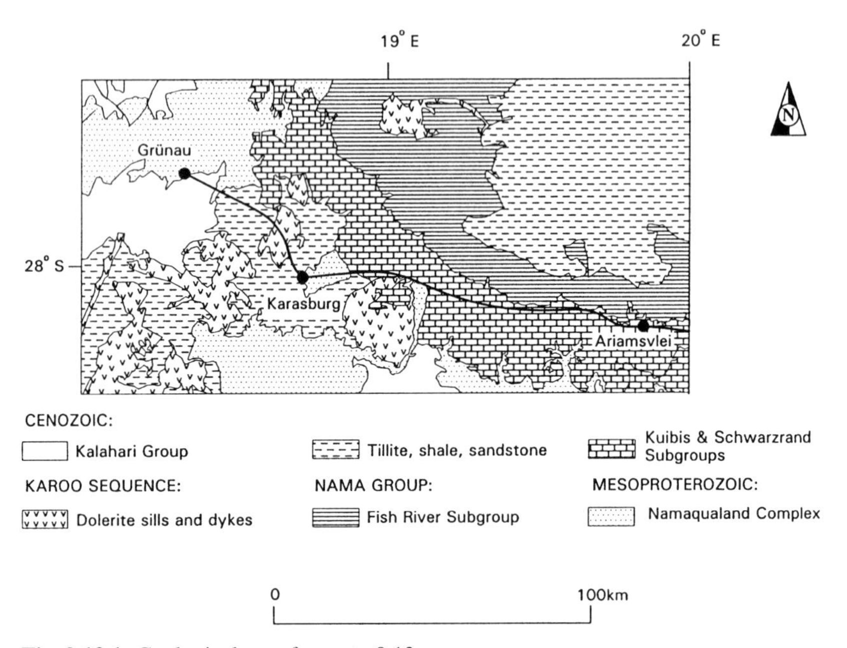


**Fig. 9.13.1:** Geological map for route 9.13.

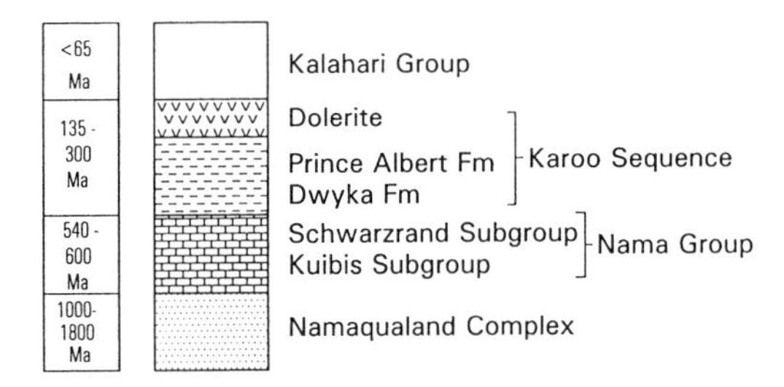


**Fig. 9.13.2:** Stratigraphic column for route 9.13.

crop. As the road descends into the valley of the Hom River, which is reached 30 km from Grünau, it moves into the stratigraphically higher shale of the Prince Albert Formation of the Karoo Sequence. The road continues in shale, and then in overlying Karoo dolerite, until it reaches Karasburg about 51 km from Grünau. There is consequently limited relief, and a corresponding lack of outcrop. The hills seen to the NE are composed of sediments of the Nama Group, while to the SW, there are some isolated hills of Karoo sediments capped by Karoo dolerite.

The town of Karasburg is underlain by migmatitic quartz-feldspar-biotite gneiss of the Namaqualand Complex. E of Karasburg, the road descends slightly and once more traverses into terrane underlain by sediments of the Dwyka Formation. The plateau seen on the southern side of the road some 25 km E of Karasburg consists of Karoo dolerites, which intruded the sediments of the Kuibis Subgroup. Some 31 km E of Karasburg, the road enters terrane underlain by these sediments, and shortly thereafter by shale and sandstone of the overlying Schwarzrand Subgroup. The surfaces on either side of the road are strewn with rubble of this material, and the road remains in these rocks up to the South African border.

The mountains of the Blydeverwacht Plateau appear in the S some 70 km E of Karasburg, and are composed of sandstone and limestone of the Kuibis Subgroup. After another 10 km, flat-topped hills of Karoo dolerite follow the road for a stretch of some 20 km, first to the S, and later to the N of the road. Thereafter, the Kainab River, a major tributary of the Orange River is crossed, and the mountains of the Platrand Plateau, composed of the same lithologies as the Blydeverwacht Plateau, become visible in the distance in the S. To the N of the road a vast peneplain is only interrupted by longitudinal red dunes of the Cenozoic Kalahari Group, which have developed on Nama and Karoo bedrock. Although Passport Control is at Ariamsvlei, the actual South African border is reached 16 km further E.

## 9.14 Keetmanshoop – Gochas – Stampriet

The 303 km long route leaves Keetmanshoop to the NE on road 29, turns W to Gochas on road C18 and follows the course of the Auob River N of Gochas along road C15 to reach Stampriet. The landscape is initially dominated by the Karoo age dolerite hills N of Keetmanshoop (see 8.5), as well as Karoo Sequence sediments, but for the larger part of the journey the road runs in an area underlain by Cenozoic sediments of the Kalahari Group, which constitute the infill of the Kalahari Basin (Figs. 9.14.1, 9.14.2).

Keetmanshoop itself is underlain by dolerites. As the road towards Gochas leaves town, the dolerite hills appear prominent in the NE. The dolerite forming these hills has an Early Jurassic age of about 180 Ma, and intruded along numerous dykes. It occurs in an area of more than 18000 $km^2$ extending for about 170 km in a N-S direction and about 110 km from E to W. Spheroidal weathering of the dolerite has led to the formation of large and small boulders strewn over the dykes and their vicinity, which frequently form distinct boulder ridges.

About 4 km from town, the road moves into terrane underlain by tillite and shale of the Permo-Carboniferous Dwyka Formation of the lower Karoo Sequence, and after another 5 km by shale and mudstone of the Prince Albert Formation, also Karoo Sequence.

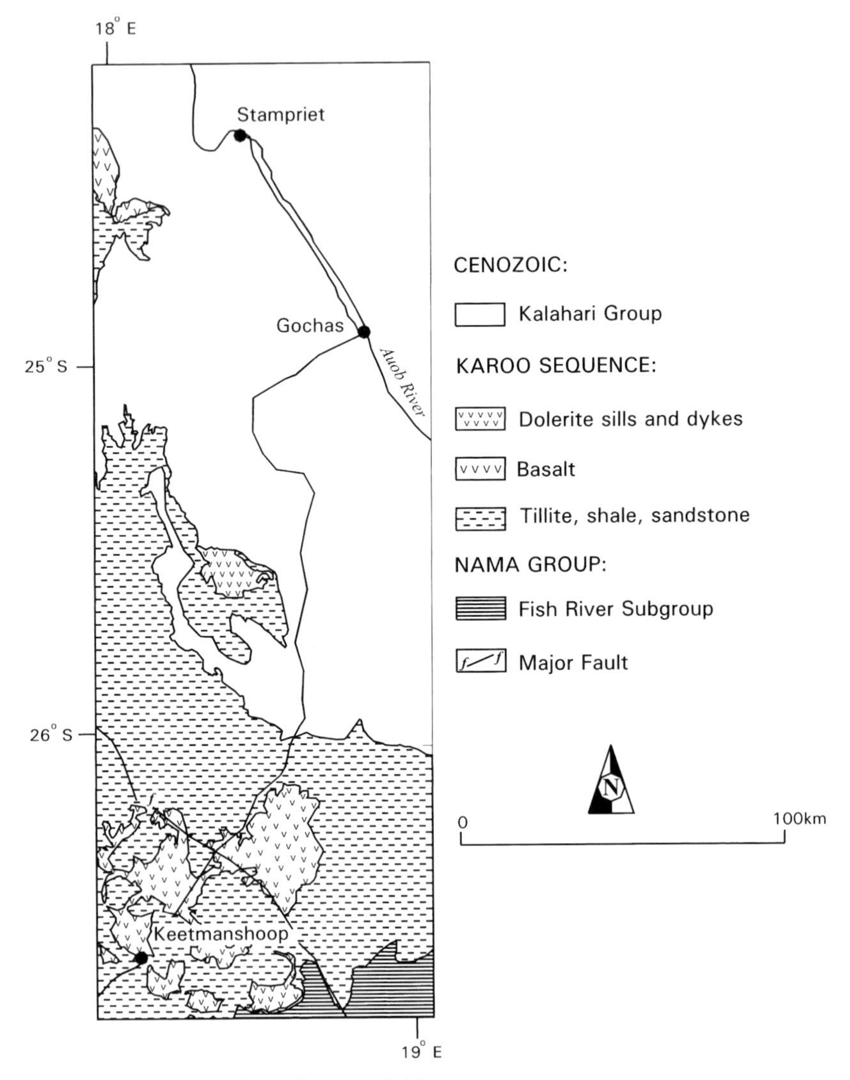


**Fig. 9.14.1:** Geological map for route 9.14.

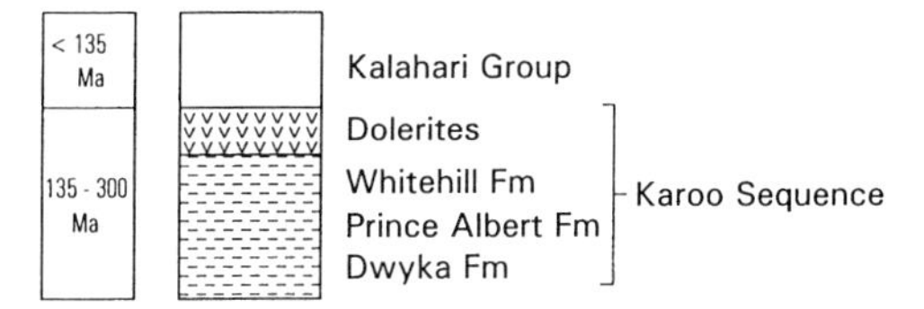


**Fig. 9.14.2:** Stratigraphic column for route 9.14.

The road then approaches higher terrane which is chiefly composed of dolerite. On the flanks of this elevated terrane are outcrops of dark carbonaceous shale of the Whitehill Formation of the upper Karoo Sequence, which on weathering attains a white colour, so giving the formation its name. After 15 km the Quiver Tree Forest on farm Gariganus 157 is reached, where, apart from the trees and dolerite boulders, well-preserved specimen of *Mesosaurus*, a fossil occurring in the Whitehill Formation (see 5.) may be observed. This lizard-like fossil has, because of a discovery in the Irati Shale of Brazil, provided early proof of the continental drift.

For the next 70 km the road remains in an area underlain by Karoo Sequence sediments extensively intruded by dolerites, and is characterized by the typical weathering of the dolerites. Thereafter it enters terrane underlain by the sediments of the Kalahari Group. These sediments are essentially flat-lying sands, and there are not many outcrops. The sand is often cemented by a siliceous calcrete, which characteristically forms in places, where evaporation exceeds precipitation, such as the Kalahari. Calcrete is a layer of hard surface limestone, which builds up by solution and re-deposition of lime by meteoric waters and often cements sand and gravel beds.

A characteristic feature of the landscape are the clay-filled pans so typical of the Kalahari. They are distinctive, Late Cenozoic to recent geomorphic elements that have formed by groundwater activity, combined with the high evaporation rate. This evaporation rate also leads to the crystallisation of appreciable amounts of salt in these pans, and salt was at one time actually produced from the Vertwall Pan, some 100 km NE of Keetmanshoop. The pans are mostly underlain and surrounded by pedogenic hardpan calcrete, but often also develop on top of Karoo dolerite.

Some 200 km NE of Keetmanshoop the C18 road turns E towards Gochas. The road now cuts across bright orange linear dunes that are as characteristic of the Kalahari as the pans. They are believed to have formed during the last glacial period some 16 000 to 20 000 years ago, when a large high-pressure cell circulated over the subcontinent, resulting in strong, cold, dry winds producing the dunes that are still seen today. The dunes carry vegetation, in some cases even large Camelthorn trees, which indicate near surface groundwater, as well as rainwater.

Gochas is reached after some 43 km, and from there the C15 road turns NW towards Stampriet. It largely follows the bed of the Auob River, a major drainage along the edge of the Kalahari Basin and towards the Orange River, which runs parallel to the dunes in an inter-dune valley. Stampriet is reached 78 km N of Gochas. The village of Stampriet is situated in the area of the Stampriet artesian aquifer, which is extensively used for irrigation purposes (see 6.). It is by far the most extensive aquifer in Namibia, and is contained in Karoo Sequence sediments with two artesian sandstone horizons, the upper Auob sandstone and the lower Nossob sandstone, sandwiched between shales and in Kalahari Group sediments forming an upper aquifer (Symons et al. 2000).

## 9.15 Mariental – Aranos – Leonardville – Gobabis

by Pete Siegfried and Gabi Schneider

The C 20 road from Mariental heads in an easterly direction towards Stampriet and Aranos. It turns off the main road B1 some 11 km N of Mariental. The road to Gobabis via Leonardville turns of to the N just before Aranos and follows the course of the Nossob River. While this route remains in Kalahari Group terrane for most of its course, it enters rocks of the Damara Sequence and Nama Group closer to Gobabis (Figs 9.15.1, 9.15.2).

At the turn-off, the area is underlain by shales of the Dwyka Formation of the Karoo Sequence, changing shortly thereafter to Kalkrand Basalt. After 10 km the road cuts through the Weissrand Escarpment, where the unconformity between the calcretised gravels of the lower Kalahari Group and the underlying Karoo rocks is exposed. The unconformity may be described as para-conformable, because, even though a long time period of some 100 Ma separates the two sequences, the layers are almost parallel and concordant to one another. The Weissrand Plateau is higher in the S and thickens to the E.

The road then crosses over a series of bright orange linear dunes. These dunes are interpreted to have formed during the last glacial period some 16 000 to 20 000 years ago. A large high-pressure cell was circulating over the subcontinent, and the resulting strong, cold, dry winds of this time produced the dunes that are still seen today. Stands of large trees, such as Camelthorn, are characteristic and indicate near surface groundwater, as well as rainwater. The Auob River is crossed twice, some 35 km and 40 km from Mariental. Stampriet is reached some 52 km E of Mariental.

The village of Stampriet is situated in the area of the Stampriet artesian aquifer, which is extensively used for irrigation purposes (see 6.). It is by far the most extensive aquifer in Namibia, and contained in Karoo Sequence sediments with two artesian sandstone horizons, the upper Auob sandstone and the lower Nossob sandstone, sandwiched between shales and in Kalahari Group sediments forming an upper aquifer (Symons et al. 2000).

The Auob River is crossed again just E of Stampriet, and as the road continues towards Aranos, the area is covered by undulating grassland with large Camelthorn trees. This landscape is fairly typical for the Kalahari and extends eastwards and towards the

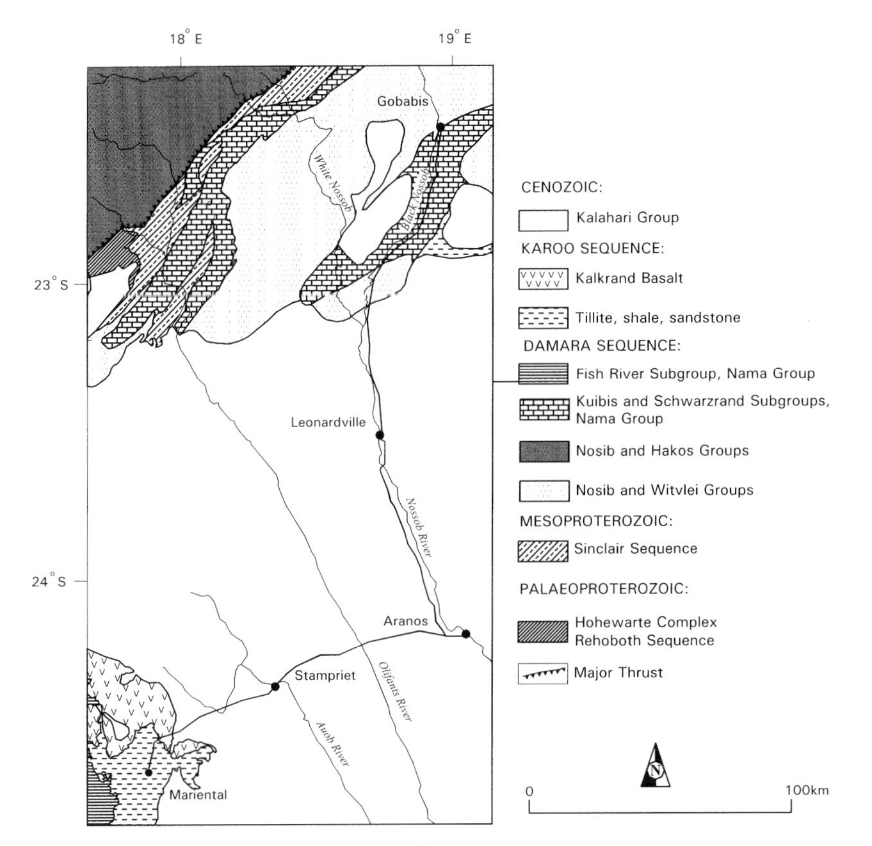


**Fig. 9.15.1:** Geological map for route 9.15.

**Fig. 9.15.2:** Stratigraphic column for route 9.15.

Botswana border and beyond. Interdune flats are usually covered in much scrubbier vegetation, with the yellow flowers of *Rhygozum* bushes a common sight. The Olifants River is crossed about 31 km E of Stampriet, and after another 40 km, Aranos is reached, with the C20 road to Gobabis turning N some 7 km before Aranos.

The road now follows the course of the Nossob River. This part of the river is regarded as fairly recent, and it is possible that this drainage, and other related drainages such as the Auob and Olifants, were formed as a response to warming and increased rainfall which followed the end of the last glacial priod some 16 000 to 20 000 years ago. Incision of these rivers was controlled by the N-S trend of the dunes resulting in the parallel directions of these drainages.

The road reaches Leonardville after some 90 km and continues northwards within the Nossob valley. It leaves this valley about 45 km N of Leonardville, and climbs up over a prominent ridge of quartzite of the Kamtsas Formation of the lower Nosib Group. This ridge marks the end of the Kalahari-age Nossob, and the river splits into the White and the Black Nossob Rivers. Scattered deposits of Dwyka diamictites of the lower Karoo Sequence in the valley floor, which have an age of 300 Ma, suggest that this part of the drainage is much older. The road returns to the valley after a further 10 km, although this time it is the Black Nossob. It remains in this valley up until Gobabis. The landscape between here and Gobabis is very flat and underlain by sandstone of the Kuibis Subgroup, Nama Group. Gobabis is reached 134 km N of Leonardville.

## 9.16 Windhoek – Dordabis – Leonardville

The first 24 km of this route follow the road from Windhoek to Gobabis (see 9.17), and then the C23 road turns off to the S towards Dordabis and Leonardville. While the first part of the journey is in terrane underlain by Palaeoproterozoic rocks and rocks of the Neoproterozoic Damara Sequence, the route between Dordabis and Leonardville comprises of Mesoproterzoic lithologies and finally ends in Cenozoic sediments of the Kalahari Basin (Figs. 9.16.1, 9.16.2).

At the C23 turn-off, the landscape is dominated by the Auas Mountains to the W and the Bismarckberge to the E (see 9.17). The road continues in an area that was affected by intense thrusting during the Damaran Orogeny, where rocks of the Hakos Group of the Damara Sequence were thrust towards the SW. For the first 5 km the road crosses a number of these thrusts, however, with little outcrop. Thereafter, it follows, for the next 8 km, a peneplain underlain by quartz-feldspar gneiss inter-layered with marble and amphibolite of the Palaeoproterozic Hohewarte Complex. The Hohewarte Complex represents one of the most ancient crustal domains in Namibia, being more than 1800 Ma old. It forms a circular dome-like structure with a diameter of approximately 40 km, and is enveloped and locally covered by rocks of Damaran age. The Hohewarte Complex has locally been intruded by the Mesoproterozoic Rietfontain Granite, which

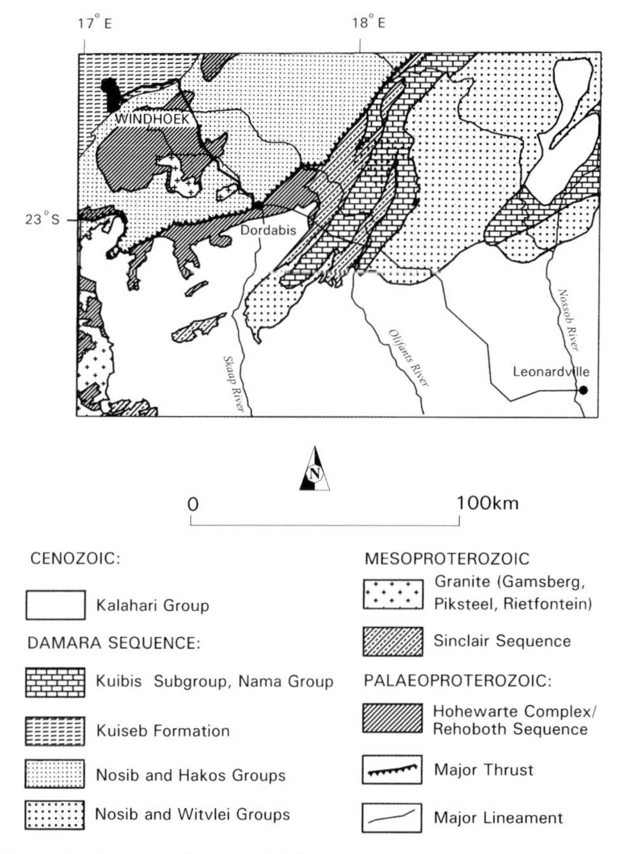


**Fig. 9.16.1:** Geological map for route 9.16.

forms hills such as the Dassiekuppe seen on the western side of the road, some 6 km from the turn-off.

The road descends into the valley of the Olifants River about 13 km S of the turn-off. This ephemeral river dewaters via the Kalahari Basin and eventually into the Orange River, and here runs parallel to the core of a major synclinal structure. The core of this structure is occupied by schist and quartzite of the Vaalgras Subgroup of the Hakos Group. They have been thrust onto schist, quartzite and marble of the Kudis Subgroup, Hakos Group. These, in turn, rest mostly conformably on the rocks of the Hohewarte

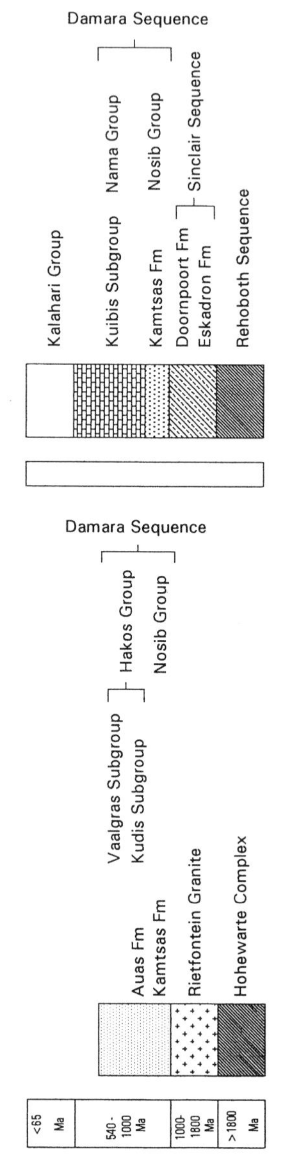


**Fig. 9.16.2:** Stratigraphic column for route 9.16.

Complex. In the E, the Rosaberg is composed of quartzite of the Auas Formation thrust onto schist and marble of the Kudis Subgroup. As the road climbs out of the Olifants River valley, it again continues in an area underlain by Hohewarte Complex lithologies.

Some 21 km from the turn-off, the road returns to terrane underlain by Kudis Subgroup lithologies. The Dreispitz Mountain towards the SE is composed of schist and quartzite of the Kudis Subgroup thrust onto the younger Vaalgras Subgroup and the Hohewarte Complex. However, after 5 km the country rocks once again belong to the Hohewarte Complex, and the landscape is flat. In the E, the Humansberg is composed of feldspathic quartzite of the Kamtsas Formation of the Nosib Group. The Kamtsas Formation was deposited in a half-graben during the early stages of rifting some 750 Ma ago. The Kamtsas Formation is here thrust towards the SE onto Hohewarte Complex lithologies.

About 34 km S of the turn-off, the road approaches a mountain range which includes the Elisenhöhe Mountain E of the road, but also extends towards the W of the road. This mountain range, which is now composed of marbles of the Kudis Subgroup with Kamtsas Formation quartzites thrust onto them, marks the end of the terrane underlain by the Hohewarte Complex. After passing this mountain range, the road continues in terrane underlain by schist of the Vaalgras Subgroup. It descends slightly into the valley of the Skaap River, but then turns E and passes between the Koedoeberg and the Hatsamas Mountains, both of which are composed of Kamtsas Formation quartzite thrust onto younger Vaalgras Subgroup lithologies.

Thereafter, the road approaches the high mountain range of the Grimmrücken with a highest elevation of 1847 m above sea level. The Grimmrücken is chiefly composed of Kamtsas Formation quartzite. The road passes a small gap in the range, passes the Skaap River marked by large Camelthorn trees, which thrive on the underground water contained in the sediments of this ephemeral river.

Dordabis is reached some 88 km from Windhoek and at Dordabis itself, the road passes a major thrust, along which quartzite of the Kamtsas Formation has been thrust onto quartzite and phyllite of the Palaeoproterozoic Rehoboth Sequence. There is, however, no outcrop, since the landscape becomes flatter as the Kalahari Basin is approached. Soon after Dordabis, the area is underlain by Cenozoic sediments of the Kalahari Group, but, some 15 km out of Dordabis, the road crosses another major fault. The country rocks are now red-brown quartzite of the Mesoproterozoic Eskadron Formation, Sinclair Sequence, and they crop out in a range of small hills seen on both sides of the road. The Eskadron Formation is host to several sediment hosted copper deposits. These copper deposits, which occur in three broad zones and are believed to have been deposited in shallow basins, have attracted extensive exploration, but to date no mining has taken place.

The road then approaches another range of hills, which are reached after 10 km. The range to the SW of the road is called Karubeamsberge, while in the NE there is the Hartebeestrückenkuppe. The road runs in a gap between the two. Again, the elevated areas are composed of quartzite of the Eskadron Fomration, while the flat area between the ranges is underlain by black limestones of the Kuibis Subgroup, Nama Group. As the road makes a slight turn to the E, it now approaches the northeastern extension of the Karubeamsberge, which here consist of metasediments of the Doornpoort Formation of the Sinclair Sequence, which overlies the Eskadron Formation. A gap in the mountain range, which is called "Doornpoort" and gave the formation its name, provides a passage for both, the road and the Olifants River, one of the major drainages into the Kalahari Basin.

The road follows the river and crosses it about 1 km after Doornpoort, and having passed Doornpoort, the road leaves behind the last rocky outcrops. It continues in an increasingly flat area, which fringes the Kalahari Basin. Although there is no outcrop, the first 20 km are underlain by quartzite of the Neoproterozoic Kamtsas Formation. Thereafter, and for the remainder of the route to Leonardville, the area is underlain by Cenozoic sediments of the Kalahari Group. About 90 km from Dordabis, bright orange, longitudinal dunes become apparent to the E of the road. These dunes are interpreted to have formed during the last glacial period some 16 000 to 20 000 years ago, when a large high-pressure cell was circulating over the sub-continent and resulted in strong, cold, dry winds producing the dunes still seen today. After another 30 km, the C48 road turns off to the E and now runs perpendicular to these prominent dunes. The Nossob River, another important drainage via the Kalahari Basin into the Orange River, is crossed after 35 km, and thereafter Leonardville is reached some 239 km from Windhoek.

## 9.17 Windhoek – Gobabis – Buitepos

by Herbert Roesener and Gabi Schneider

The journey from Windhoek to Gobabis and the Botswana border at Buitepos starts in the central Namibian highlands and ends on the western fringe of the Kalahari Basin (see 8.13). The traverse crosses the Southern Margin of the Damara Orogen, then passes into Mesoproterozoic rocks, before it continues in early Damaran rocks and then for a short while after Gobabis, in an area underlain by the Nama Group (Figs. 9.17.1, 9.17.2).

Leaving Windhoek eastwards on road B6, the geology consists mainly of quartzites of the Kleine Kuppe Formation and the Wasserberg Member, which are intercalated with the schists of the Kuiseb Formation. About 12 km from Windhoek, the turn-off to Otjihase Mine is reached. Copper is mined here from a massive sulphide ore body. The ore forming process was associated with the submarine volcanic activity that formed the Matchless Amphibolite Belt. This belt is a linear feature up to 3 km wide and some 350 km long. It consists of amphibolites, which form lenses and layers interbedded with the Kuiseb Formation schist. These amphibolites have a MORB geochemical signature, and for this reason have been interpreted as metamorphosed, syn-sedimentary submarine volcanics emplaced into sediments covering a mid-ocean ridge that had de-

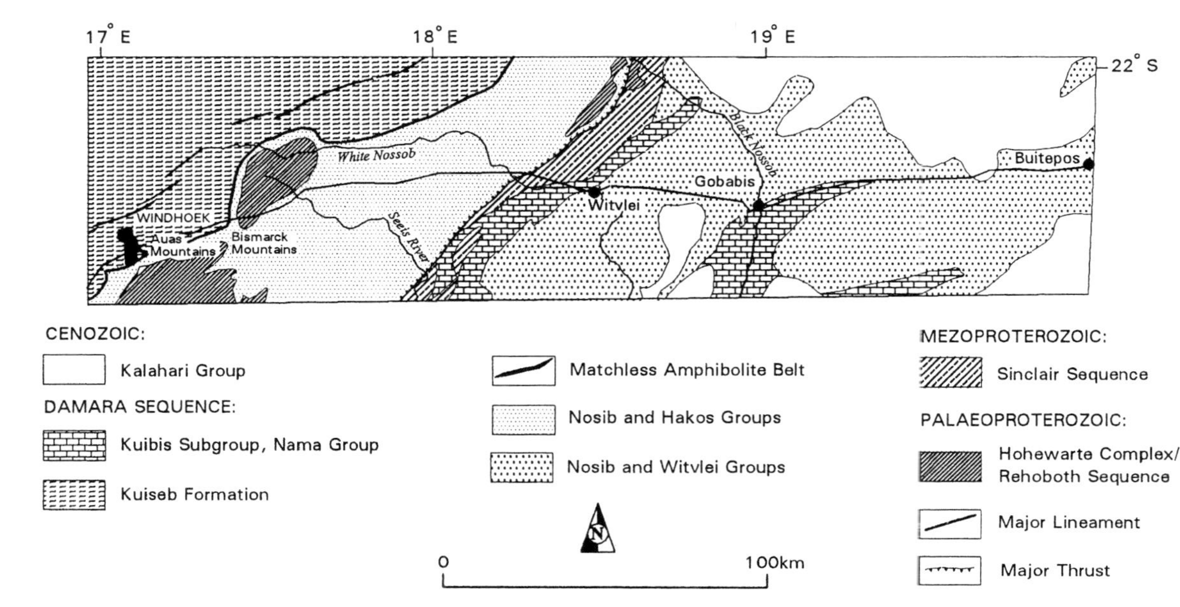


**Fig. 9.17.1:** Geological map for route 9.17.

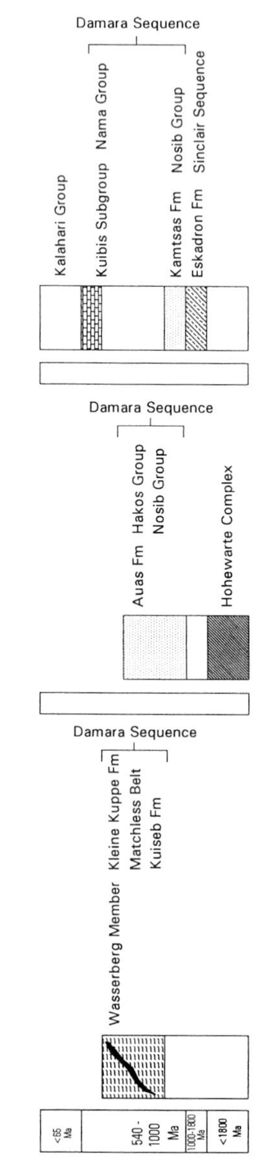


**Fig. 9.17.2:** Stratigraphic column for route 9.17.

veloped in the Damaran Southern Zone ocean. The Otjihase Mine came into production in 1975, and has since produced more than 200 000 t of copper.

Just after Kapps Farm, some 19 km E of Windhoek, the road climbs a ridge underlain by a series of distinct micaceous quartzite beds of the Kleine Kuppe Formation, which are well exposed in road cuts. From the Dordabis turn-off to about 3 km before the Hosea Kutako International Airport, the road crosses a series of thrusts in schist and quartzite of the Hakos Group. The thrust direction is towards the SW and the intense thrusting is a result of the compressive forces which were active during the Damaran Orogeny.

The schist and quartzite form the extension of the prominent Auas Mountains S of the road. These approximately 730 Ma old sediments of the Auas Formation were deposited in a spreading ocean on the continental shelf of a passive margin. They underwent subsequent metamorphism during the Damaran Orogeny, when the intensely folded rocks were eventually thrust towards the SW. They form an impressive mountain range with some of the highest peaks in the country, and Moltkeblick, at 2479 m above sea level, is the second highest mountain in Namibia. At the summits, quartzites, which owe their preservation to a high weathering resistance, terminate in steep cliffs facing S.

The wide peneplain, on which the Hosea Kutako International Airport is situated, is underlain by the Hohewarte Metamorphic Complex consisting of mica schist, porphyroblastic ortho- and para-gneiss, migmatite, granite gneiss and amphibolite. The Hohewarte Metamorphic Complex represents one of the most ancient crustal domains in Namibia, and is more than 1800 Ma old. The prominent mountains S of the airport are called Bismarckberge and consist of quartzite of the Auas Formation thrust onto Nosib Group schist.

Having reached the peneplain, and thereby the highest point on the way to Gobabis, the road has crossed one of the major water divides in central Namibia. While the area towards the NW dewaters into the Atlantic drainage, the rivers E from here contribute towards the Kalahari and Orange River drainage, and the first major river, the Seeis River, is crossed some 55 km east of Windhoek.

Further E the low lying areas are covered by Cenozoic sediments of the Kalahari Group and the ridges consist of more resistant Nosib quartzite, which can be seen in occasional roadcuts. Geological mapping in the Omitara area has revealed tight NE trending folds and thrusts with large displacements and several deformation events have been recognised. Soon after the Windhoek – Gobabis district boundary, the geology changes into lithologies of the Kamtsas Formation of the Nosib Group. The Kamtsas Formation was deposited in a southern half-graben during the early stages of rifting some 750 Ma ago, and consists of quartzites with characteristic heavy mineral layers, conglomerate bands and scattered pebbles.

The quartzites persist laterally for about 10 km, thereafter they are thrust over the Mesoproterozoic, red-brown quartzites of the Eskadron Formation, Sinclair Sequence, which is well exposed in road cuts. The Eskadron Formation is host to several sediment hosted copper deposits. These copper deposits, which occur in three broad zones and are believed to have been deposited in shallow basins, have attracted extensive exploration, but to date no mining has taken place. The White Nossob, a major ephemeral river in eastern Namibia, is crossed some 28 km before Witvlei.

The village of Witvlei is situated in a gap between a ridge formed by sandstone Kuibis Subgroup of the lower Nama Group and quartzite of the Nosib Group. The road from Witvlei to Gobabis and further on to Buitepos at the Botswana Border is mainly underlain by feldspathic quartzites of the Kamtsas Formation. These quartzites are pebbly in places and have intercalated red shale horizons towards the E.

The Black Nossob River is crossed just before Gobabis, and the town itself is situated on the contact of Nosib and Nama Group sediments. The isolated low hills in the surroundings of the town comprise Nosib Group lithologies, with Nama Group sediments forming a syncline and the road eastwards continues along the northern limb of this syncline. For the first 45 km towards Buitepos, the road climbs to just over 1500 m and overlooks the wide valley of the Chapmans River, which is named Okwa in Botswana, to the S. Likewise, to the N there is a view over the valley of the Rietfonteinriver. Further on, the landscape is characterised by the slow descent into the Kalahari Basin, with little morphological features and no outcrop near Buitepos. Buitepos is reached 298 km E of Windhoek, in an area which is underlain by Cenozoic sediments of the Kalahari Group.

## 9.18 Windhoek – Steinhausen – Summerdown – Okahandja

This 403 km long route provides a good overview of the eastern central highlands of Namibia and remains for most of its length within the Southern Zone of the Damara Orogen. It also touches the Southern Margin Zone and the western fringes of the Kalahari Basin, and towards the end crosses the Okahandja Lineament to continue in the Central Zone of the Damara Orogen (Figs.9.18.1, 9.18.2).

The route follows the B16 road for the first 16 km until Kapp's Farm, and the geology is described in 9.17. At Kapp's Farm, road 53 turns off towards the NE, in an area underlain by schist with intercalated quartzite of the Kuiseb Formation, Damara Sequence. The Kuiseb Formation is an almost 1000 m thick sequence of mica-schist, quartzite and meta-greywacke, formed from sediments deposited in a spreading ocean, that have undergone metamorphism during the Damaran Orogeny. The mountains to the N, such as Ludwigkop, Lydiakop and Neudammkuppe are also composed of rocks of the Kuiseb Formation, and are the result of extensive block faulting during the Middle Tertiary as a result of extensional tectonics.

The road continues in undulating landscape underlain by rocks of the Kuiseb Formation with small valleys that have been eroded by rivers contributing to the headwaters of the White Nossob in the E, which is one of the major drainages via the Kalahari Basin into the Orange River. The Bismarckberge are prominent in the SE, here

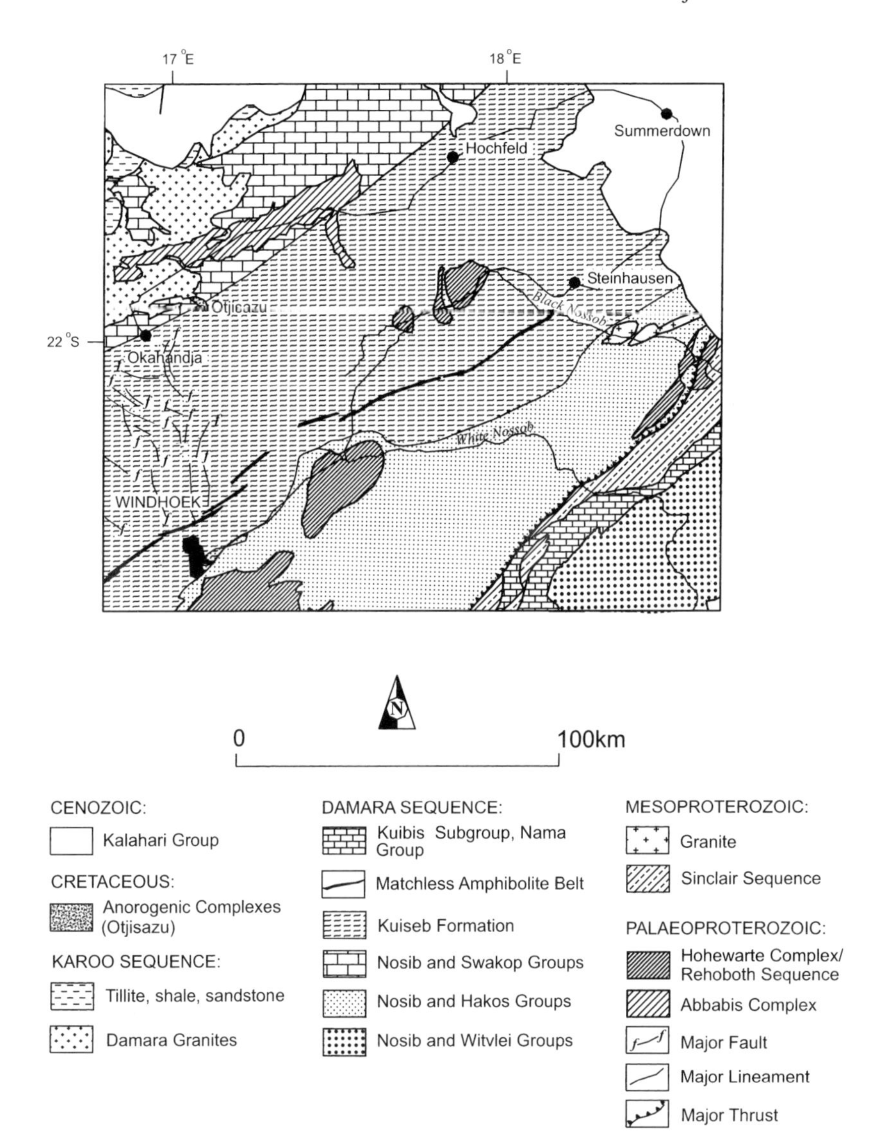


**Fig. 9.18.1:** Geological map for route 9.18.

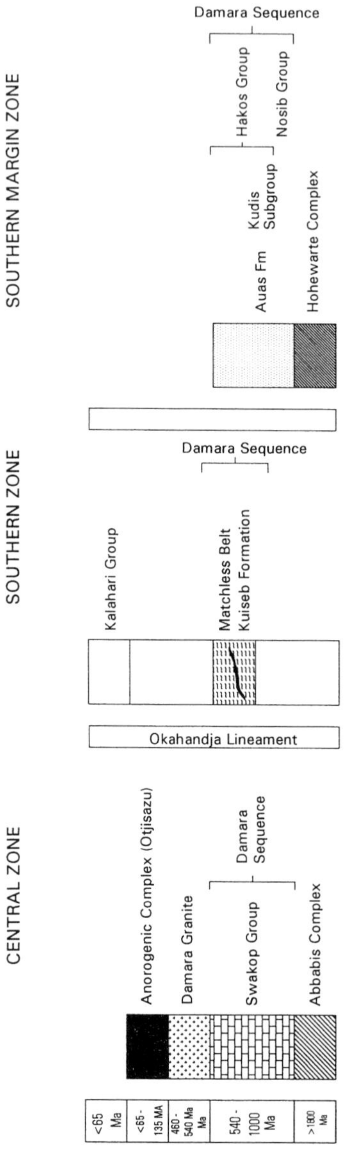


**Fig. 9.18.2:** Stratigraphic column for route 9.18.

quartzite of the Auas Formation has been trust onto schist of the Nosib Group during extensive thrusting related to the continental collision of the Damara Orogen. About 30 km from the turn-off, the road enters this zone of extensive thrusting, and is now underlain by graphitic schist of the Kudis Subgroup, Hakos Group. Shortly thereafter, however, it moves into terrane underlain by gneisses of the Hohewarte Complex, which, with an age of more than 1800 Ma, is one of the most ancient crustal domains in Namibia. The road remains in this terrane for about 11 km, then crosses the thrusted graphitic schist once more to return to terrane underlain by the Kuiseb Formation. Despite these changes in the geology, there is little difference in the landscape.

The Matchless Amphibolite Belt is crossed after another 9 km, but here there is little outcrop. The Matchless Amphibolite belt is a linear feature up to 3 km wide and some 350 km long. It consists of amphibolites, which form lenses and layers inter-bedded with the Kuiseb schist. The amphibolites have a MORB geochemical signature, and have for this reason have been interpreted as metamorphosed syn-sedimentary submarine volcanics, emplaced into sediments covering a mid-ocean ridge that had developed in the Damaran Southern Zone ocean. The amphibolites are associated with important massive sulphide deposits such as the Otjihase Mine E of Windhoek.

The road then rises slightly in schist of the Kuiseb Formation, and after 11 km reaches Onganja, to the W of which the old Onganja or Otjizonjati Mine is located (see 7.1). This mine produced some 3 000 t of ore grading 20 % copper between 1903 and 1912. It is interesting to record that the deposit had been known to the indigenous people for many years, who traveled for more than 500 km from the N to mine the ore up to a depth of 7 m (Bürg 1942). The mine was not very profitable however, as the ore had to be transported for more than 100 km by oxwagon to the nearest railway station, and from there another 300 km by rail before being shipped to Germany. After 1912, the mine was only operated intermittently, and closed permanently in 1974. Today, the old shafts and tunnels still remain. The hydrothermal ore body consists of quartz veins carrying chalcopyrite, chalcocite, cuprite, molybdenite and pyrite, and Onganja is famous for the largest known euhedral cuprite crystals in the World measuring up to 14 cm in length and width and weighing more than 2 kg (see 4.). They can be seen in many museums around the globe.

Onganjaberg, composed of schist of the Kuiseb Formation is prominent in the N. As the road continues the landscape flattens and outcrops become rare, although the road eventually descends into the valley of the Black Nossob, which, together with the White Nossob becomes the larger Nossob River, one of the more important ephemeral rivers of eastern Namibia. The road then follows the course of this river for 15 km before reaching Steinhausen. After 15 km and again after 30 km respectively, the road passes areas, which are underlain by gneisses of the Hohewarte Complex. At the boundary of the farms Omitiomire Nord and Otjere, some 36 km from Steinhausen, alpine-type serpentinite is developed along the contact of the Hohewarte Complex and the enveloping Damaran schist. The serpentinite is part of a number of serpentinite bodies, which occur in a belt close to the tectonic boundary between the Southern Zone and the Southern Margin Zone of the Damara Orogen. The serpentinite forms pods and lenses. Steinhausen is reached 186 km from Windhoek.

After Steinhausen, the road descends gradually into the Kalahari Basin (see 8.13), and the landscape becomes very flat and is underlain by the Cenozoic sediments of the Kalahari Group. Summerdown, 49 km from Steinhausen, is located on the western fringes of the Kalahari Basin, and is also underlain by these sediments.

From Summerdown, the C31 road heads westwards towards Hochfeld and Okahandja. Initially, it remains in an area underlain by Kalahari Group sediments, but about 30 km before reaching Hochfeld it rises out of the Kalahari Basin, and continues in an area underlain by Kuiseb Formation schist. This is, however, hardly noticeable, and there are no outcrops. Hochfeld is reached some 61 km W of Summerdown.

For the first 40 km after Hochfeld the road remains in terrane underlain by schist of the Kuiseb Formation with basically no outcrop. It then crosses the Okahandja Lineament, which separates the Central Zone of the Damara Orogen from the Southern Zone. It marks major changes in the stratigraphic succession, structural style and age of metamorphic events. The lineament trends NE and represents a deep penetrating zone of weakness in the Earth's crust which has repeatedly been active and has had a major influence throughout the depositional and tectonic history of the Damara Orogen (Miller 1983). The lineament is only noticeable in the field by virtue of the fact that N of it, gneisses of the Mesoproterozoic Abbabis Complex, the basement to the Damaran rocks in the area, occur, which form some hills S of the road.

The road remains in terrane underlain by these gneisses of the Abbabis Complex, and there is a slight increase in the relief of the landscape. Some 55 km after crossing the Okahandja Lineament, the prominent Otjisazu Mountain, which is a Cretaceous anorogenic complex comprising of granite, carbonatite and pyroxenite, can be seen in the far S, and from here the country rocks change to Damaran granite for the greater part remainder of the route to Okahandja. The change is marked by the appearance of numerous granitic hills, displaying typical onionskin weathering. About 107 km from Hochfeld the road joins the main road from Okahandja to Otjiwarongo (see 9.20), some 8 km N of Okahandja. After 2 km on the main road, the road enters a terrane where Damaran granites form a series of small hills with schist and marble of the Nosib and Swakop Groups underlying the broad valleys between them. The Okakango River, a tributary of the Swakop River, is crossed just N of Okahandja.

## 9.19 Okahandja – Otjiwarongo – Otavi – Tsumeb

by Herbert Roesener, Gabi Schneider and Volker Petzel

The road northwards from Okahandja to Otjiwarongo is about 177 km long. It commences in Damaran granites in the Okahandja area, traverses through Karoo Age rocks before moving back into Damaran age rocks close to Otjiwarongo, from where it remains in Damaran sediments until reaching Otavi after 119 km and Tsumeb after another 63 km (Figs. 9.19.1, 9.19.2)

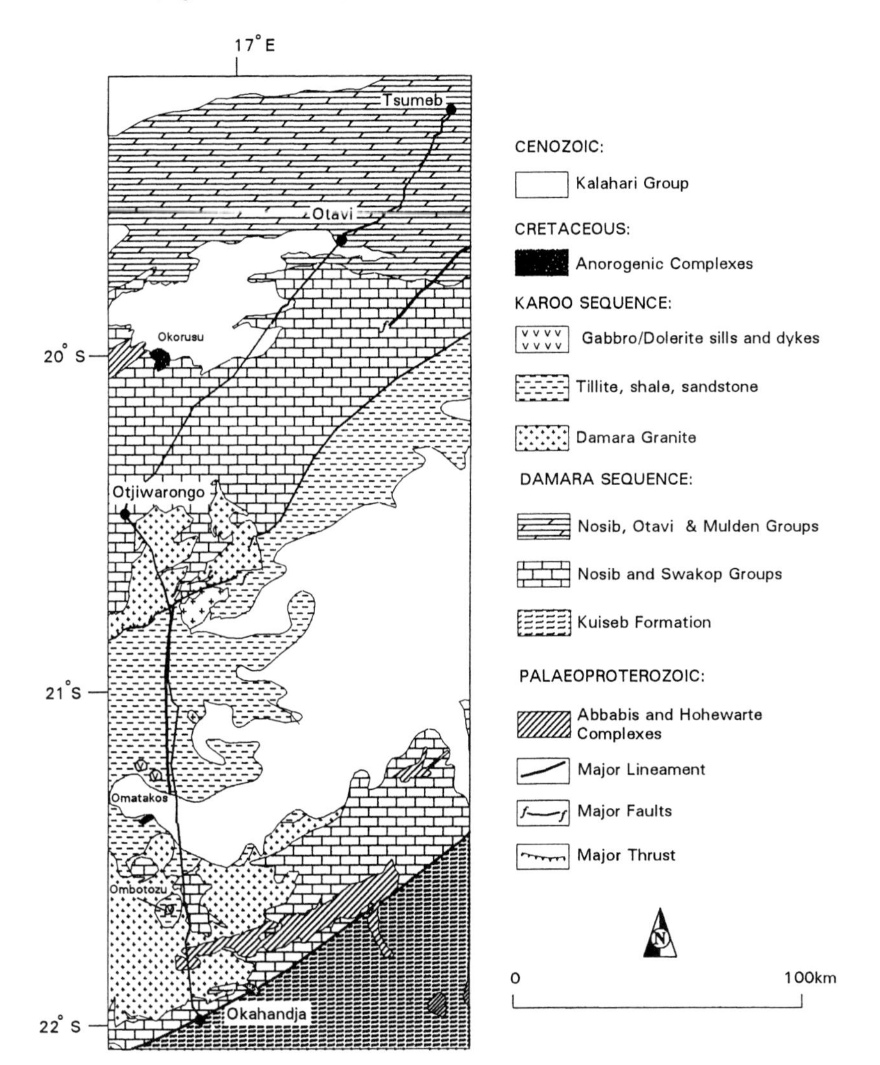


**Fig. 9.19.1:** Geological map for route 9.19.

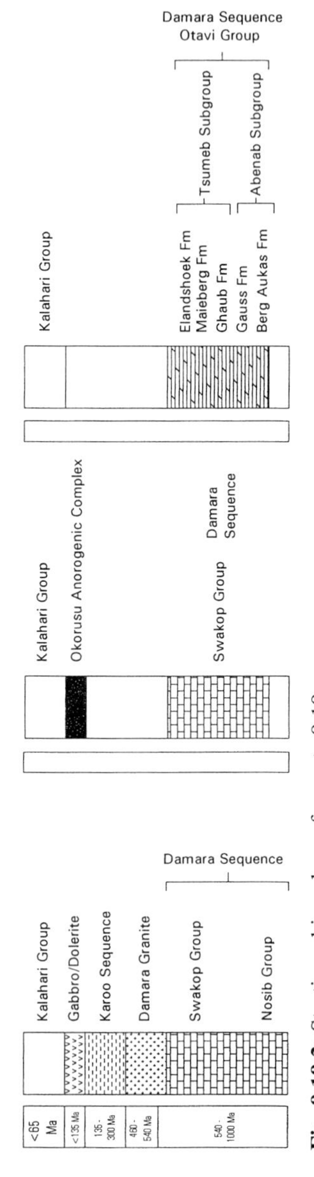


**Fig. 9.19.2:** Stratigraphic column for route 9.19.

N of Okahandja, schist and marble of the Nosib and Swakop Groups of the Central Zone of the Damara Orogen are found, intruded by granites. The Okakango River, a tributary of the Swakop River, is crossed just N of Okahandja, whereafter Damaran granites form a series of small hills with schist and marble of the Nosib and Swakop Groups underlying the broad valleys between them.

Further to the N, scattered isolated inselbergs of Karoo age rocks form prominent topographic features. The first of these inselbergs, Ombutozu Mountain, becomes visible some 6 km outside Okahandja, and is passed on the western side of the road about 40 km N of the town. Ombutozu Mountain rises about 350 m from the surrounding plain and has a height of 1916 m above sea level. It forms an irregularly embayed table mountain, capped by an extensive sill of olivine gabbro. The gabbro, which is some 100 m thick, rests on 140 m of Karoo sediments, including light coloured fine-grained feldspatic sandstone, and baked, reddish violet mudstone, with intercalations of laminated shales and grayish violet clayey sandstones. A 15 – 20 m thick sill of decomposed gabbro and a lower greenish mudstone some 15 m thick constitute the lower portions of the hill, which is scree covered at the base. A line of low hills trending N is present on the northern side of the mountain. It is believed that these hills represent the fissure connecting the gabbro of the Ombutozu Mountain with that of the Omatakos (see 8.23).

The twin peaked Omatako Mountains are located some 90 km N of Okahandja, and are first seen on clear days some 40 km from Okahandja. From here to the Omatakos, the route continues in terrane underlain by sediments of the Karoo Sequence, and in places the road is slightly elevated, providing a magnificent view into the endlessly flat Kalahari Basin (see 8.13). Immediately to the W of the road another Karoo inselberg, Okandumondjuwo Mountain, can be seen and just before the Omatako Mountains are reached, Klein Omatako Mountain, also a Karoo inselberg, occurs directly W of the road. Here, the road is again slightly elevated, providing an even better view of the Kalahari Basin in the east. The hill seen in the NE is composed of Damaran granite.

The road crosses the Omatako River 92 km N of Okahandja. The river has its headwaters near the Omatako Mountains only 260 km from the sea in the W. However, upwarping of the continent edge related to the break-up of Gondwanaland and the formation of the Great Escarpment and the Kalahari Basin, created a watershed causing this river to flow northeastwards and today, the Omatako River drains into the Kavango River E of Nyangana, some 700 km NE of the Omatako Mountains.

About 97 km from Okahandja, Mount Etjo (see 8.19) becomes prominently visible in the W, followed by the Okonjima Mountain after some 126 km. Like Mount Etjo, Okonjima forms a table-topped range capped by hard Etjo Sandstone. The Waterberg (see 8.32) appears in the NE after a further 8 km and in this area the Waterberg Basin has developed above metamorphic basement rocks of the Damara Orogen. Schists and granitic gneisses of the Damara Sequence form an uneven basement floor to the Karoo Sequence and this has resulted in several sub-basins. The basin is bounded in the N by the prominent NE-trending Waterberg Fault, which the road crosses within another kilometer, about 40 km S of Otjiwarongo. Along the fault older Damaran rocks were thrust in a southeasterly direction onto younger Karoo age rocks during Jurassic times.

The road to Waterberg is passed after a further 12 km, and from here to Otjiwarongo the terrane is underlain by Damaran granite, which also forms the typical round granite monoliths outcropping in this area. However, 6 km outside Otjiwarongo, the road crosses a ridge consisting of marble, schist and calc-silicate of the Swakop Group, before it moves once more back into granite, which also underlies the town. The ridge continues to the SW, where it is composed of sediments of the Nosib and Swakop Groups.

A few kilometers after leaving Otjiwarongo for Otavi on the B1, the road enters terrane underlain by the Northern Zone of the Damara Orogen. A number of hills and ridges consisting of Swakop Group marble appear on both sides of the road and form prominent landmarks. In the E, the Honigberg rises above the surrounding low-lying areas, while a little further the Jägershof Mountain, which is part of a NE-striking marble range, soars to an impressive 1656 m to the W of the road, and still further, the marbles of the 1822 m high Orosberg Range can be seen to the E in the distance.

The road to Okorusu Mine turns off to the NW about 40 km north of Otjiwarongo. The mine is one of the chief producers of acid-grade fluorspar in the World. The fluorspar is associated with a carbonatite, which is part of the post-Karoo anorogenic complex forming Okorusu Mountain, which is visible in the distance. Since its first opening in the 1920s, the mine has produced over 550 000 t of fluorspar concentrate of various grades. Trucks convey the concentrate to the Okorusu railway siding next to the road, from where it is railed to Walvis Bay for shipping.

Before passing from the Northern Zone of the Damara Orogen, with its associated amphibolite facies metamorphism, into the region of the Northern Platform with Otavi Group carbonates and more moderate greenschist facies metamorphism, another prominent Swakop Group Formation marble ridge, the Okumukanti range, can be seen in the distance to the SE of the road.

As one approaches Otavi, the elongated Elefantenberg, consisting of diamictites, iron-formation and dolomites of the Abenab Subgroup and lower Tsumeb Subgroup, Otavi Group, lies to the E of the road. Its dark bulk can be seen from quite a distance and contrasts with the mountains to the N of it, which consist of lighter carbonates of the Tsumeb Subgroup.

The road then continues in a very flat landscape with no outcrop, underlain by rocks of the Swakop Group, until, with no visible difference in the landscape, the underlying lithologies change to rocks of the Otavi Group about 5 km outside Otavi. Otavi is reached 119 km N of Otjiwarongo.

Departing Otavi towards the N, the northern limb of the Otavi valley syncline, which is composed of Maieberg and lower Elandshoek Formation dolomite, lies to the E. In the distance to the W, the platform-like Kudib Mountain consists of Elandshoek Formation dolomite, and still further W the Langeberg Range can be seen. When the road makes a wide turn to the N, about 23 km from Otavi, small outcrops of Ghaub Formation diamictite, which forms the base of the Tsumeb Subgroup, can be seen to the E of the road. Further back, Nosib quartzites form another series of low hills, while to the W, layered Maieberg Formation limestone and dolomite form the prominent Schumanstal Range.

About 38 km from Otavi, the road cuts through a small outcrop of Berg Aukas Formation limestone and Gauss Formation dolomite. Further N, a succession of anticlines and synclines is crossed. Here the center of a flat calcrete and sand-covered area is formed by Nosib quartzites, while the low reddish hills of quartzite to the E of the road indicate the anticlinal hinge zone, and the higher, lighter-coloured hills beyond consist of Abenab carbonates.

Before Tsumeb, the "Ten Mile Pass" has to be negotiated, where hills formed by Elandshoek Formation dolomites can be seen on both sides of the road. Along the pass, they are exposed in road cuts, while the low-lying area below the pass is underlain by Abenab rocks. Tsumeb is reached 63 km N of Otavi. This well known mining town owes its existence to the Tsumeb copper-lead-zinc deposit, which was mined since 1906. The mine closed in 1998, during its life of 92 years it had produced more than 27 million tons of ore yielding approximately 1,9 million tons of copper, 3 million tons of lead and 1 million tons of zinc, as well as numerous by-products. It is a unique deposit, and with over 230 different minerals from two oxidations zones, is world-famous. Today the upper levels remain open and are exclusively mined for mineral specimens (see 4.).

## 9.20 Omaruru – Kalkfeld – Otjiwarongo

by Volker Petzel

Between Omaruru and Otjiwarongo, along the C33 road, the geological scenario encompasses sedimentary and intrusive rocks belonging to the Damara and Karoo Sequences, thus spanning more than 400 Ma of earth history (Figs. 9.20.1, 9.20.2). S of the Omaruru River to the right, the Damara-age granite pinnacle of the Omaruru Koppie forms a distinctive landmark, and like many of the granites and pegmatites of this area, hosts a small occurrence of tourmaline.

On leaving Omaruru, to the E of the road, a flat-topped mountain range, formed by almost horizontally bedded clastic sediments of the Omingonde Formation, Karoo Sequence, is visible on the horizon, while several hillocks of porphyritic Salem granite of Damaran age rise above the plain in the middle distance.

A little further on, to the NW, a couple of distinctive ranges can be seen, which consist of marbles of the Damara-age Karibib Formation. The nearer Epako range, together with the Elefantenberg on the far left, is formed by massive calcitic marble of the Arises River Member. It overlies the banded marbles of the Otjongeama Member of the more distant, higher Tjirundo range. A good view of the banded and coarse-grained

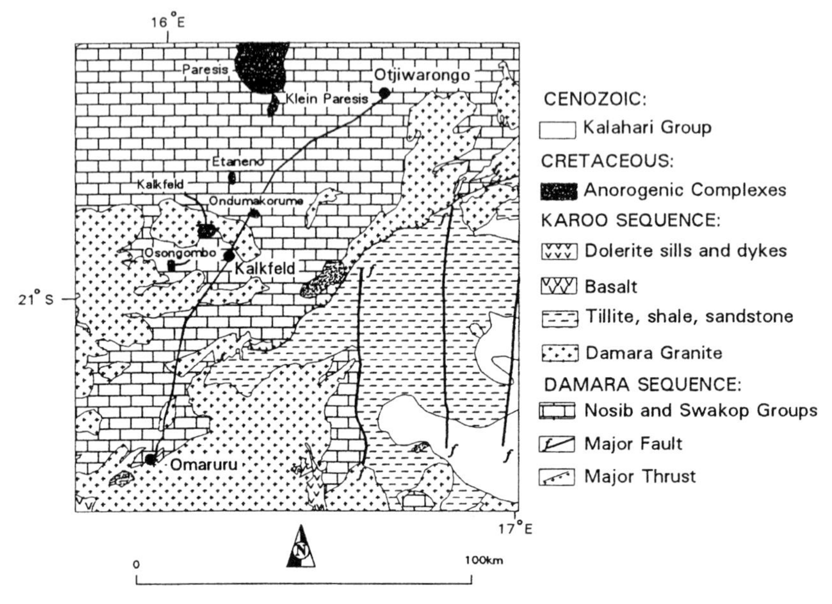


**Fig. 9.20.1:** Geological map for route 9.20.

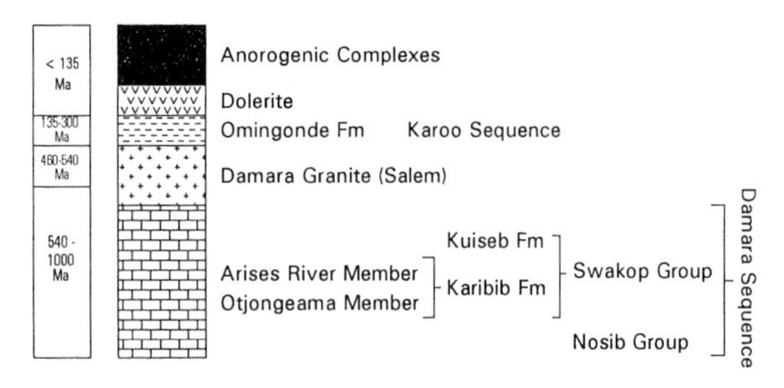


**Fig. 9.20.2:** Stratigraphic column for route 9.20.

massive marbles is provided when looking towards the SW from the railway bridge ca. 27 km outside Omaruru. Alluvial gold was produced in small quantities between 1937 and 1943 on the Farm Epako, which gave the range its name.

Approximately 9 km after leaving Omaruru, a sharp-crested post-Karoo olivine-dolerite dyke, which runs parallel to the road for several kilometers, forms dark hillocks to the E. It belongs to the Cretaceous swarm of dykes, which follows the Omaruru Lineament, a zone of crustal weakness that existed from Damara to post-Karoo times. The dykes cut older sedimentary and intrusive Damara rocks, and in this area are related to the Erongo ring dyke (see 8.7). Looking SE from the railway bridge, one has a good view of the dyke, as well as of the previously mentioned Omaruru Koppie.

On the farm Otjua, about 32 km from Omaruru, Damara marble is intruded by Damara granite resulting in the formation of tungsten-bearing skarns. Some 2 km further on, the double peak of the Okongue Koppie can be seen to the SW, this hill, like other less prominent but also distinctive domes in this area, consists of Damara leucogranite, which intruded the surrounding Damara metasediments.

In a road cut, about 37 km after leaving Omaruru, porphyritic Salem granite of Damara-age outcrops and is cut by a dolerite dyke showing distinctive blocky weathering. Within the granite remnants of light-coloured skarn are also discernible. A little further, another prominent granite outcrop with skarn formation and a dolerite intrusion can be seen in the SE, while the largely flat country to the NW is underlain by Kuiseb schist, Swakop Group, Damara Sequence.

Damara marble again outcrops about 49 km from Omaruru to the left of the road, and the next few kilometers are characterized by a complex amalgamation of Damara metasediments and granites, and, locally, post-Karoo dykes. While the granites form distinctive koppies or domes, low hills near the road consist of skarn. Intermingled granites, skarns and marbles occur in several road cuts, the latter often consisting of large calcite crystals due to re-crystallization during contact metamorphism.

The granites usually display characteristic granite-type weathering and one particularly excellent example can be seen to the SE of the road, approximately 60 km from Omaruru, where Salem granite with dark biotite and large pink feldspar crystals contains light-grey and dark xenoliths of the country rock in various stages of assimilation. Skarnification occurs in a millimeter thick band along the contact between grey coarse-grained marble and the crosscutting granite. Both the marble and granite have in turn been cut almost at right angles by a 2 m thick dolerite dyke. On the last leg to Kalkfeld typical granite domes are a frequent feature of the landscape on both sides of the road.

In Kalkfeld the road passes the turn-off to the Dinosaur tracks at Otjihaenamaparero (see 8.25). After leaving the town, road C33 continues over Damara metasediments and granites, with a low range of Damara granites to the SE. Stopping at the Outjo turn-off one has a good view of several anorogenic complexes of Cretaceous age. To the SW the Kalkfeld Carbonatite Complex forms a prominent landmark in the middle distance, while the pyramid-shaped Etaneno Complex lies a few kilometers to the W and the sharp-peaked Ondumakorume Complex lies to the E, next to the road. To the N in the distance, the jagged crest of Klein Paresis and the massive Groot Paresis Complex, with the highest peak at 1806 m above sea level, can be seen (see 9.24).

NW of Etaneno a low range of hills consisting of Nosib Group quartzite is visible, and continuing towards Otjiwarongo, isolated koppies of Damara granite are once again in evidence on both sides of the road. Ahead to the SE, the Ohiwa Mountains are formed by a larger intrusion of Damara granite and further along the road, the Nosib Group quartzites of the Good Hope range with a peak of 1773 m at Alter Römer Berg come into view to the NE. These are replaced, just before Otjiwarongo, by numerous pinnacles of Damara granite, which rise above the flat landscape in the far distance. In contrast to the more gently rounded quartzite hills, they have relatively sharp peaks or domes, while the ridge supporting a telecommunications tower consists of Damara marble, distinguished by its lighter colour.

## 9.21 Otavi – Grootfontein – Tsumkwe

by Gabi Schneider and Volker Petzel

This route passes the Otavi Mountainland consisting of Proterozoic dolomites and limestones to reach Grootfontein after 95 km, from where it continues to Tsumkwe for 276 km in an area underlain by Cenozoic rocks of the Kalahari Group (Figs. 9.21.1, 9.21.2).

The road leaves Otavi towards the NE in an area underlain by shales and quartzites of the Ghaub Formation of the Otavi Group, with little or no outcrop. The landscape is dominated, however, by the fast approaching high rising Otavi Mountainland to the N and NW. About 7 km outside Otavi, on the northern side of the road, a huge quarry has been opened up in the overlying Tsumeb Subgroup dolomites, which has been used for railroad ballast. At this locality, the road passes close to the hinge zone of the Otavi Valley syncline, which gives rise to the 1812 m peak to the S of the road, and on the hillside the distinctive eastward-dipping bedding within the carbonates is clearly visible through the sparse vegetation of thornbush and grass.

After another kilometer, the road passes the synclinal hinge zone through a fairly narrow gorge which leads into the Otavi Valley, and good outcrops of Tsumeb Subgroup dolomites occur on either side of the road in road cuts. The valley then opens up, with the high hills of the Otavi Mountainland rising up to 1800 m above sea level on both sides of the valley. The entire Otavi Group is present in the Otavi Valley which forms the core of the 50 km long, E-W trending, northward overfolded Otavi Valley syncline, with dolomites and limestones of the Abenab Subgroup, overlain by dolomites and limestones of the Tsumeb Subgroup on the flanks and the younger phyllite of the Kombat Formation of the Mulden Group, partly covered by sand and calcrete in the core of the syncline (Deane 1933). The road follows this synclinal structure and along the lower slopes the mountains consist of dolomites of the Hüttenberg Formation, whilst the hills themselves are formed by the bedded and massive dolomites of the Elandshoek Formation. The low, dark-coloured ridge, that extends between the northern valley flank and the road, and which can best be seen near Kombat, consists of oolites and stromatolites (see 5.) of the uppermost Hüttenberg Formation.

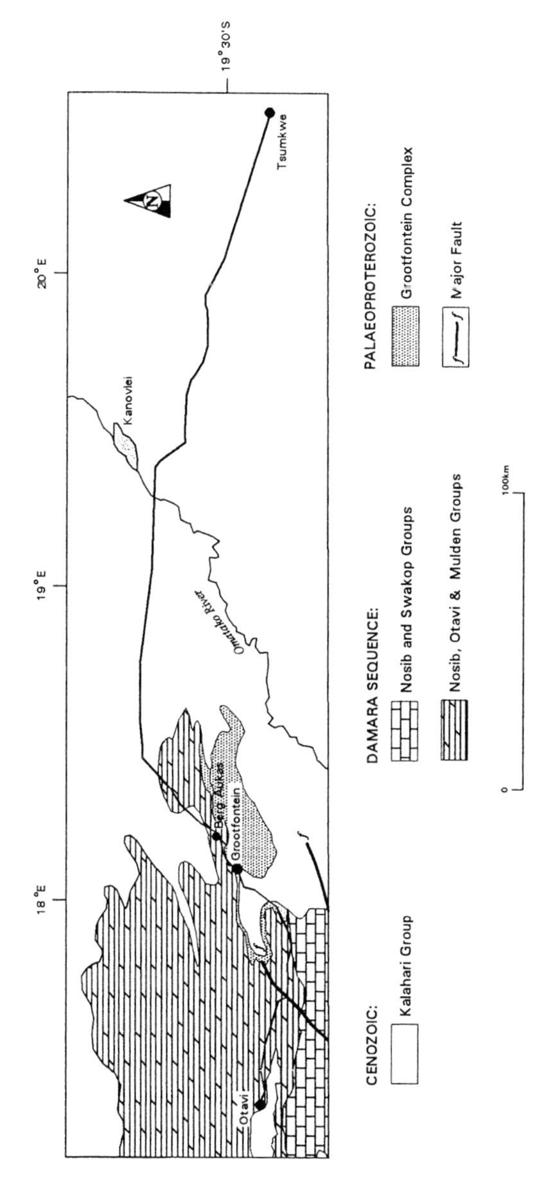


**Fig. 9.21.1:** Geological map for route 9.21.

**Fig. 9.21.2:** Stratigraphic column for route 9.21.

Some limited accumulations of fine-grained sediments of the Kalahari Group also occur within the Otavi Valley, and 13 km from Otavi, brickworks on the southern side of the road make use of such alluvial clays together with weathered Kombat Formation phyllites. The copper deposits in the Otavi Valley are, however, of much greater significance. Mining has been conducted here since prehistoric times, and exploration, mainly for base metals has been proliferous in these mountains since the end of the 19th century. Some of the defunct small mines and prospects, such as Baltika Mine, Gross Otavi Mine and Wolkenhaube Prospect can still be seen from the road. A little further on from the brickworks, for example, extensive scree slopes on the northern hillside indicate the position of the abandoned Baltika Mine, where up until 1942 vanadium was mined from fissures and karst-fill sandstone, mostly in open cuts.

The Kombat Mine, 43 km from Otavi, is to this day a significant producer of copper and lead. The orebody consists of galena, pyrite, chalcopyrite, bornite and chalcocite associated with tectonic and sedimentary breccias, dolomite and arkose. The mine has been productive intermittendly since 1911, during which time well over 10 million t of ore have been produced.

The most prominent rock type outcropping along the road through the Otavi Valley is the intensely silicified dolomite of the lower Elandshoek Formation, which is weathered to produce numerous characteristic pinnacles. This feature can be observed wherever this unit is present, but it is particularly well-developed some few kilometers after leaving Kombat and also along the first kilometer of the turn-off track towards the Hoba Meteorite (see 8.12), some 15 km from Kombat, where the rocks outcrop directly beside the road.

The Rodgerberg Mine is another defunct mine which can be seen some 4 km down the track to the meteorite, although it is only recognized by scree slopes on the hillside to the left. The mine produced copper concentrate and the first dioptase discovered in Namibia came from the Rodgerberg and the adjacent Guchab mines. Still on that track, after it makes a sharp turn, dolomites of the Gauss Formation of the lower Otavi Group, which form the southern flank of the Otjihaenena Valley anticline, can be seen on the right and the hills on the far side of the valley also consist of Gauss dolomites, with the

Berg Aukas Formation outcropping at the base. The core of the anticline is here underlain by highly metamorphic rocks of the approximately 1800 Ma old Grootfontein basement complex. They are only preserved locally along the valley flanks, discordantly overlain by the much younger Abenab carbonates, but are mostly concealed by white calcrete and Kalahari sand.

The agricultural potential of the Otavi Valley is another important feature of the region. Fertile soils have developed from the phyllites of the Mulden Group, which, combined with the high rainfall usually experienced in this region, makes the Otavi Valley one of the few areas in Namibia supporting large scale crop production.

Approaching Grootfontein, the road leaves the Otavi Valley and the Kalahari Basin (see 8.13) becomes visible after some 73 km from Otavi, but the mountains of the Otavi Mountainland remain prominent in the W. Grootfontein, situated on the northwestern edge of the Kalahari Basin, is reached after another 22 km. Although the area surrounding Grootfontein is underlain at depth by the Grootfontein Complex, outcrops are confined to the overlying calcrete of the Kalahari Group. This calcrete was formed by evaporation of near-surface groundwater, which carried calcium carbonate derived from the surrounding Otavi Group limestones. Although the maximum annual rainfall of only 750 mm is high for Namibian circumstances, near-surface waters are no longer abundant, which is also on account of the karst topography. The development of calcrete therefore suggests a more humid climate in the recent geological past. The calcrete is well exposed in some road cuts on the northern end of town. Many of the buildings in Grootfontein have also used calcrete as a building material.

From Grootfontein, the road follows a northeasterly direction along the edge of the Kalahari Basin. The huge basin is well visible to the E of the road, especially some 5 km N of Grootfontein, where calcrete has formed a small rise. Calcrete for road building is quarried to the W of here.

Some 13 km out of Grootfontein, a few isolated inselbergs of Otavi Group dolomite become visible in the E, while generally, the landscape becomes rapidly very flat. The turn-off to the old Berg Aukas Mine is passed 16 km N of Grootfontein. The mine is prominently visible in the E, located on one of the inselbergs. The vanadium-lead-zinc deposit was discovered in 1913 and mined intermittendly until 1978, during the period 1962 to 1978 alone some 2,7 million t of ore were produced, which contained 1,2% vanadium oxide, 5 % lead and 22 % zinc. The galena-sphalerite ore contains secondary descloizite which occurs as lenses in the dolomite. Descloizite specimens from this locality are world famous, and the largest vanadinite crystal in the World, measuring 12 cm in length, was found at the nearby Abenab Mine (see 4.). Berg Aukas is also famous for the first Miocene hominoid discovery south of the equator, *Otavipithecus namibiensis* (see 5.). Both, the vanadinite and some fossilised remains of the hominoid are displayed in the museum of the Geological Survey of Namibia in Windhoek.

The occasional dolomite and limestone inselberg can still be seen in the E up to 37 km N of Grootfontein, thereafter the landscape shows little or no relief, and is underlain entirely by the Cenozoic sediments of the Kalahari Group. About 51 km N of Grootfontein, road C44 to Tsumkwe turns off to the E. The entire 225 km to Tsumkwe remain in this terrane with little morphological expression. Some 89 km from the turn-off, the Omuramba Omatako, which forms a large vlei to the N of the road, is crossed. This major drainage has its headwaters near the Omatako Mountains between Okahandja and Otjiwarongo. However, the inland drainage resulting from isostatic movements related to the break-up of Gondwanaland and the formation of the Great Escarpment and the Kalahari Basin, has created a watershed which caused this river to flow northeastwards to the Kavango River. Today, the Omatako River drains into the Kavango River east of Nyangana, almost 700 km NE of its source, which is only 260 km from the Atlantic Ocean.

Tsumkwe, which is reached 136 km after the Omatako crossing, is located well within the western reaches of the Kalahari Basin. There are no rock outcrops and there are few morphological features in the Tsumkwe area, other than the typical undulating Kalahari landscape.

## 9.22 Grootfontein – Rundu – Katima Mulilo – Ngoma

by Pete Siegfried and Gabi Schneider

This 830 km long route follows the northwestern edge of the Kalahari Basin between Grootfontein and Rundu, before leading in an easterly direction into the Kalahari (Figs. 9.22.1, 9.22.2). The geology along the route illustrates some important aspects of the on-going deposition of sediments within this central southern African inland basin (see 8.13).

The area around Grootfontein is situated on the northwestern edge of the Kalahari plain. This plain is underlain by calcrete of the Kalahari Group, which in turn is underlain by Damaran dolomites and limestones of the Otavi Group, and also by the Groot-

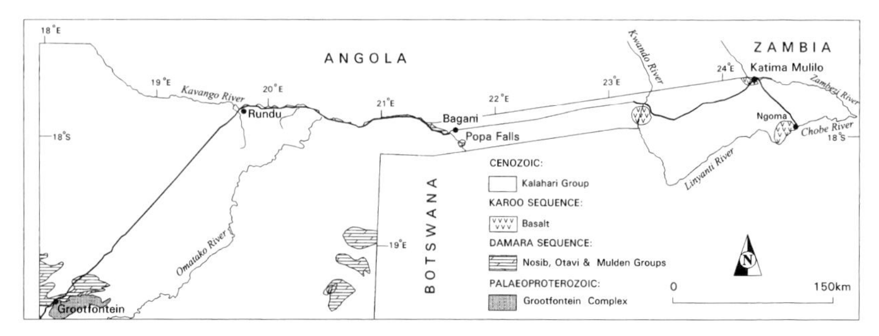


**Fig. 9.22.1:** Geological map for route 9.22.

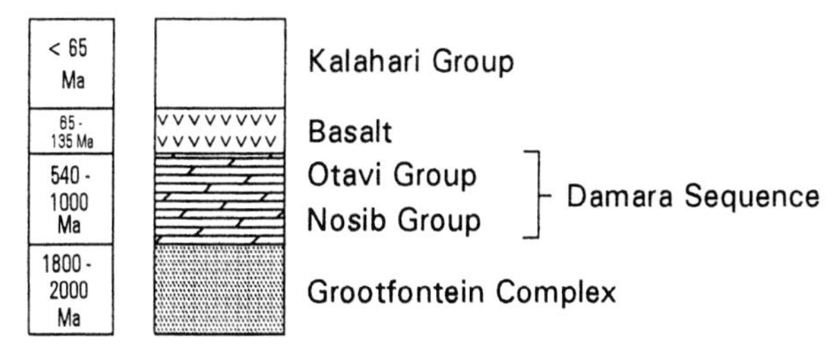


**Fig. 9.22.2:** Stratigraphic column for route 9.22.

fontein Complex in the area around Grootfontein. Throughout the plain, however, outcrops are confined to the calcrete of the Kalahari Group. The calcrete is formed by the evaporation of near-surface groundwater, which contained calcium carbonate derived from the surrounding Otavi Group limestones. Although the maximum annual rainfall of 750 mm is high for Namibia, near-surface waters are no longer abundant, partly on account of the karst topography. The development of calcrete therefore suggests a more humid climate in the recent geological past. The calcrete is well exposed in some road cuts on the northern end of town. Many of the buildings in Grootfotein have also used calcrete as a building material.

After leaving Grootfontein, the road follows in a northeasterly direction along the edge of the Kalahari Basin, the huge expanse of which is well visible to the E of the road. This panorama is particularly visible some 5 km N of Grootfontein, where calcrete has formed a slight rise elevation. Calcrete used for road building is quarried to the W of the road.

Some 13 km from Grootfontein, a few isolated inselbergs of Otavi Group dolomite become visible in the E, while generally, the landscape becomes increasingly very flat. The turn-off to the old Berg Aukas Mine is passed 16 km N of Grootfontein and the mine is clearly visible on one of the inselbergs in the E. The vanadium-lead-zinc deposit was discovered in 1913 and was intermittently operational until 1978. During the period 1962 to 1978 some 2,7 million t of ore were produced, containing 1,2 % vanadium oxide, 5 % lead and 22 % zinc. The galena-sphalerite ore also contains secondary descloizite which occurs as lenses in the dolomite. Descloizite specimens from this locality are world famous, and the largest vanadinite crystal in the world, measuring 12 cm in length, was found at the nearby Abenab Mine (see 4.). Berg Aukas is also famous for the first find of a Miocene hominoid south of the equator, *Otavipithecus namibiensis* (see 5.). Both, the vanadinite and some fossilised remains of the hominoid are displayed in the museum of the Geological Survey of Namibia in Windhoek.

The occasional dolomite and limestone inselberg can be seen in the E up to 37 km N of Grootfontein, thereafter the landscape shows no more relief, and is underlain entirely by Cenozoic sediments of the Kalahari Group. About 70 km N of Grootfontein, after the turn-off to Tsumkwe, the abundance of scattered Makalani palms *(Hyphanae petersiana)* indicates the presence of shallow groundwater in the sediments of the Kalahari Group. Shortly thereafter, one becomes aware of prominent E-W trending longitudinal dunes composed of whitish sand. These dunes were formed during the last glacial period some 16 000 to 20 000 years ago. A large high-pressure cell was circulating over the subcontinent and the resulting strong, cold, dry winds of this time produced the dunes seen today. The area is known as the Mangetti, due to the abundance of Mangetti trees *(Sclerocarya birrea)*, which commonly occur along the dune crests.

A large river valley is crossed approximately 5 km before Rundu is reached some 248 km from Grootfontein. This valley is the Omuramba Ndonga, and is a palaeo-channel of the Omuramba Omatako, which rises near the Omatako Mountains between Okahandja and Otjiwarongo (see 8.23). The inland drainage resulted from isostatic movements following the break-up of Gondwanaland which caused this river to flow northeastwards to the Kavango River, although its headwaters are only 260 km from the sea in the W. Today, the Omuramba Omatako drains into the Kavango River not far E of Nyangana.

The landscape between Rundu and Bagani is relatively flat, broken only when the road crosses some tributaries of the Kavango River. The Omuramba Omatako, mentioned above, some 60 km E of Rundu, is one, and another is the very prominent Omuramba Katere, some 88 km E of Rundu.

The turn-off to Popa Falls and Bagani is reached 212 km E of Rundu, and here, S of the road, some very minor rocky outcrops of folded quartzite with intercalated limestone, which may be correlated with the Damaran Nosib Group (Hegenberger 1987) are exposed. The same rocks are found in the rapids at Popa Falls, and the area represents an isolated outcrop of Damaran age rocks within the Kalahari plain. After a further kilometer, the Kavango River, which in Botswana forms the world famous Okavango Delta, is crossed by a bridge.

There are no outcrops on the road from Bagani through the Western Caprivi, although there is a noticeable change in the colour of the soil from whitish to reddish as the bridge over the Kwando River at Kongola is approached. This is caused by the area in the E being partly underlain by Karoo age basalts with associated lateritic weathering. Just E of the Kongola bridge, a hill located immediately to the S of the road, is in part composed of such highly altered basalt. The Kwando River, a highly meandering river feeds the Linyanti swamps to the S along the border with Botswana, and is crossed some 410 km from Rundu.

The Kavango and Kwando are interior drainage systems, which terminate in the Okavango Delta and the Linyanti swamps respectively, and are controlled by rifting which has produced faults on a regional scale. These regional faults follow a NE-SW direction, and are associated with local perpendicular faults. The courses of the Kavango and the Kwando Rivers follow these faults, and the Okavango and Linyanti swamps are bounded by such faults in the SE. The course of the Chobe River in the eastern part of the Caprivi is also controlled by similar faults. Both, Karoo basalts and Kalahari Group

**Fig. 9.22.3:** The Katima Mulilo rapids in the Zambesi River.

sediments have been uplifted by the faults, which indicates that they are of a young geological age. The faulting is interpreted to be an extension of the East African Rift System, and continuing earthquake activity supports this interpretation (Reeves, 1972, Hegenberger 1987, Scholz 1976).

The next 103 km leading to Katima Mulilo again follow an area underlain by Cenozoic sediments of the Kalahari Group, and there is absolutely no rock outcrop. Katima Mulilo itself is situated on the banks of the Zambesi River, just downstream from some major rapids (Fig. 9.22.3), which are caused by outcrops of Karoo basalt. There are again no rock outcrops, as the road continues towards Ngoma, and the Botswana border at Ngoma is marked by the course of the Chobe River. Across the river on the Botswana side, Karoo basalts, uplifted along a fault controlling the course of the Chobe River, can be seen. These basalts are, however, once again highly altered.

## 9.23 Grootfontein – Tsumeb – Oshakati – Ruacana

by Volker Petzel and Gabi Schneider

This almost 500 km long route starts off in rocks of the Damara Sequence, but soon leads into the younger Kalahari Group sediments of northern Namibia. Only just before Ruacana, Damaran rocks are encountered once again (Figs. 9.23.1, 9.23.2).

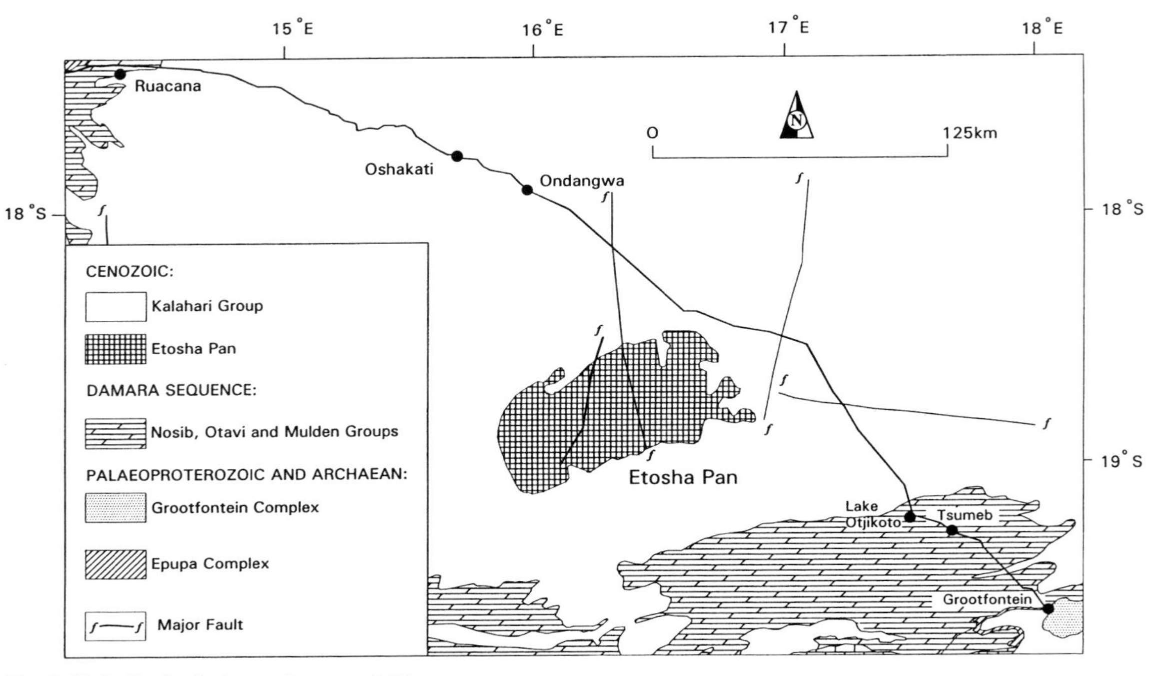


**Fig. 9.23.1:** Geological map for route 9.23.

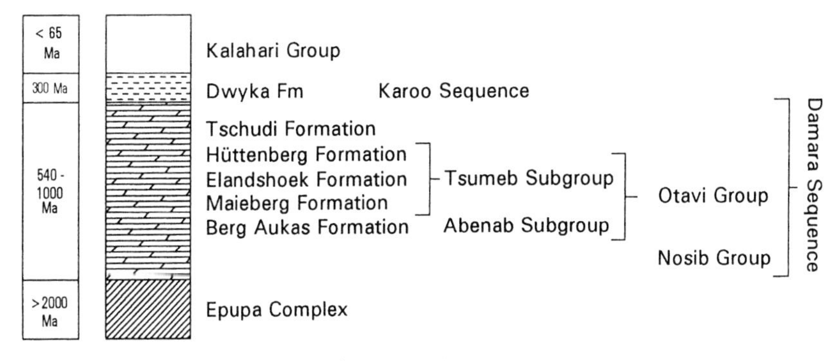


**Fig. 9.23.2:** Stratigraphic column for route 9.23.

As the town of Grootfontein lies on the edge of the Kalahari Basin, only few outcrops are present and the calcrete-covered country is generally flat. On the way to Tsumeb, road C42 crosses several synclinal and anticlinal axes, thus traversing repeatedly Otavi Group stratigraphy, which is, however, mostly covered by sand and calcrete. The first elevation visible from the road in the NE is the Jägersquell Mountain, which consists of silicified dolomites of the Elandshoek Formation, while a little further on a ridge approaching the road from the W consists of Berg Aukas Formation dolomites. After the road makes a sharp turn to the E it continues between the older Abenab Subgroup to the S and the younger Tsumeb Subgroup in the N, before turning once again northwards traversing the Tsumeb Subgroup rocks. A road cutting shows thinly bedded limestones and phyllites as well as dolomites of the Maieberg Formation, which are succeeded by the dolomites and subordinate limestones of the Elandshoek and Hüttenberg Formations. The silicified dolomites of the lower Elandshoek Formation are once again crossed at the turn-off to the Khusib Springs copper mine.

The hills visible approximately 20 km from Tsumeb and again, just before the town, also consist of Elandshoek and Hüttenberg Formation carbonates, but except for sporadic road cuts outcrops near the road are sparse. The road leading into town just after the cross-roads passes through the upper part of the Tsumeb Subgroup and the Tsumeb Valley itself is underlain by sandstones of the Tschudi Formation, which outcrop sporadically along its flanks.

From Tsumeb, the B1 road continues in a northwesterly direction in an area underlain by carbonates of the Hüttenberg and Elandshoek Formations, however, outcrops are rare, although about 10 km from Tsumeb, there is a quarry where the carbonate is crushed to produce various sizes of aggregate for building material. Lake Otjikoto (see 8.17), a deep sinkhole in dolomites of the Maieberg Formation is reached after another 10 km. Thereafter, no more outcrops in Tsumeb Subgroup lithologies are found.

Just before Otjikoto, the Tsumeb Geophysical Research Station is passed on the northern side of the road, although it is somewhat concealed by the low thorn bush which thrives in the area. The station is operated by the Geological Survey of Namibia and monitors seismological (earthquake) activity as part of an international network of seismic stations, secular variation of the Earth's magnetic field and cosmic emissions. The station was established in the 1950s by the Max Planck Institute of Germany to conduct ionospheric research. Although this research was discontinued several decades ago, some of the high antenna masts remain.

As the landscape becomes more flat, the road now approaches the western part of the Kalahari Basin (see 8.13). The area is underlain by sand, clay and calcrete of the Kalahari Group, which form the cover rocks to the Damaran rocks of the Owambo Basin. This huge sedimentary basin, which underlies the whole of central northern Namibia, contains sediments to a thickness of up to 7000 m, with the uppermost unit, the Kalahari Group, reportedly attaining a maximum thickness of 400 m.

Some 140 km N-NW of Tsumeb, Etosha Pan (see 8.8) can be seen prominently towards the S. The morphology here is characterized by the slight relief of parallel, SE trending dunes. The area is part of the Cuvelai drainage system, which developed at the end of the Pliocene, after the Kunene River ceased to supply Etosha with water. This drainage system consists of numerous seasonal rivers, named oshanas. Oshanas carry water only during seasonal floods called efundjas, and occupy the interdune valleys. Flooding usually occurs in February, and turns the area, which is otherwise characterized by sand and sparse vegetation, into a swampy garden.

Ondangwa is reached 247 km NW of Tsumeb, and Oshakati after a further 35 km. The road then continues for another 180 km with little or no rock outcrop, before it enters an area underlain by rocks of the Nosib Group, Damara Sequence, some 15 km before Ruacana. A good section through the Nosib Sediments is exposed along the road to the Ruacana Falls. The upper layers are well bedded and the lower units area composed of a basal conglomerate that contains clasts several tens of centimeters in size. In addition to the Nosib sediments, the younger, fluvio-glacial sediments of the Dwyka Formation, Karoo Sequence, crop out in places along the road to the hydro-electric power station.

The Ruacana Falls are situated west of Ruacana at the contact of gneisses of the Epupa Complex and the sediments of the Nosib Group, which are more amenable to erosion. The falls are 700 m wide with a total drop of 124 m, and the river has cut its way northwards forming a 120 m deep zig zag gorge, with almost vertical walls. The gorge is some 1,6 km in length and follows the contact of the gneiss with the sediments. The valley widens rapidly from a narrow point some 600 m upstream the falls forming numerous channels inter-fingering with islands of Epupa gneiss. During the rainy season, when the river is in full flood, the falls are continuous, but as the discharge decreases significantly during the winter months, the fall becomes a series of individual falls. Downstream from the falls fluvio-glacial sediments of the Dwyka Formation, Karoo Sequence, are exposed, indicative that the Kunene River valley was partly shaped by a huge glacier during the Dwyka glaciation some 300 Ma ago. See also Excursion 9.25.

The Calueque Dam, some 12 km upstream of Ruacana, regulates the influx of water to Ruacana, and the diversion weir at Ruacana Falls channels the water to the underground turbines of the 240 MW Ruacana powerstation. This station produces up to 60 % of Namibia's electricity when the Kunene is in flood, but on average it provides some 20 % of Namibia's annual power consumption.

## 9.24 Otjiwarongo – Outjo – Okaukuejo

by Herbert Roesener and Gabi Schneider

This 190 km long route starts in schist and carbonate rocks of the Swakop Group, Damara Sequence, passes the sediments of the Northern Platform with inliers of Palaeoproterozoic rocks, before ending in the Cenozoic sediments filling the northern Namibian Owambo Basin (Figs. 9.24.1, 9.24.2). The C38 road from Otjiwarongo follows a northwesterly direction towards Outjo. The larger part of the road is underlain

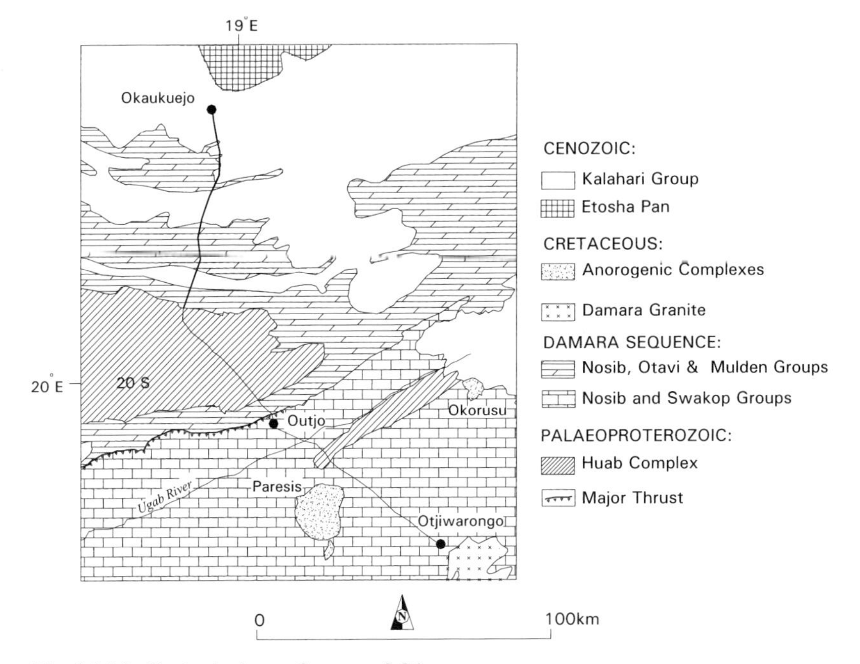


**Fig. 9.24.1:** Geological map for route 9.24.

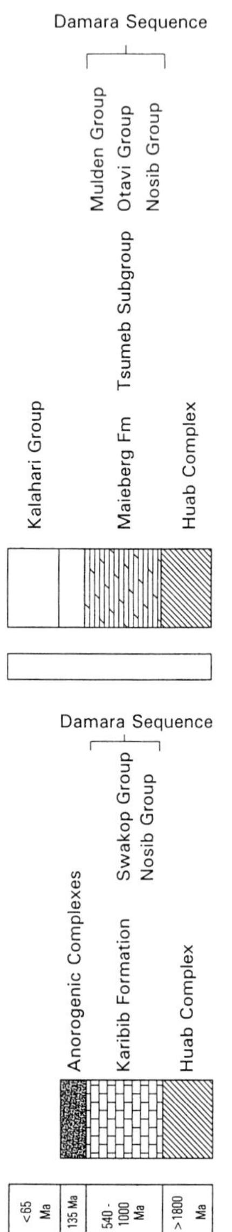


**Fig. 9.24.2:** Stratigraphic column for route 9.24.

by Swakop Group schist and marble of the Northern Zone of the Damara Orogen. These rocks are folded with the broad fold axes trending NE, which gives rise to a slightly undulating topography.

On leaving Otjiwarongo, there are some hills to the SW of the road which consist of quartzite and schist of the Nosib Group. Thereafter, the Klein and Gross Paresis Mountains form a prominent ridge to the W of the road. The rocks of these anorogenic complexes are of post-Karoo Age, and are part of the magmatism related to the break-up of Gondwana. The Paresis Complex is made up of alkaline volcanic and sub-volcanic rocks. The majority of the exposed rocks are pyroclastic flows of rhyolitic composition with basaltic lavas and gabbros present as minor components. The sub-volcanic rocks are syenite, bostonite, micro-granite and various feldspathoidal rocks, with each of these petrological units representing an individual volcanic event that originated from different centers within the complex. Finally, the volcano collapsed, forming a caldera and this has resulted in the inward dip of the lithologies.

About 50 km from Otjiwarongo, the road crosses a broad synform of gneiss of the Palaeoproterozoic Huab Complex bordered by Nosib Group quartzite and schist and Karibib Formation marble. The hills to the NE of the road are composed of Karibib marble. Just before Outjo, near the turn-off to Kalkfeld, Cenozoic sediments form prominent terraces SW of the road. The terraces consist of conglomerates cemented by limonitic calcrete and were deposited by the Ugab River. The Ugab River is one of the larger ephemeral rivers in Namibia and has its headwaters between Outjo and Otavi (see 8.31). These conglomerates can be seen in road cuts on the southern fringes of town.

A sudden change in the landscape also occurs just before Outjo is reached, where hills composed of limestone showing karstification mark the transition from the Northern Zone of the Damara Orogen to the the Northern Platform. The road from S of Outjo, through town and out towards Okaukuejo passes an interesting cross-section through folded Otavi and Mulden Group rocks of the Northern Platform. The first prominent ridge about 1 km S of Outjo consists of laminated limestones of the Maieberg Formation. The outcrop forms the southern limb of a syncline, the northern limb of which underlies the town of Outjo. The area immediately NW of Outjo is occupied by an anticline with older phyllitic schist of the Nosib Group in its core. The next two ridges, 1,5 km and 5 km N of Outjo, consist of Tsumeb Subgroup dolomite which is exposed in road cuts, and forms the limbs of a syncline. The core of the syncline consists of phyllite of the Mulden Group. The folding in this area is due to NW-SE directed compressional stress that resulted in the formation of major open folds with NE-trending fold axes, as well as second generation tighter parasitic folds (Thirion 1970).

N of Outjo, for the first 9 km, the landscape is dominated by undulating hills with a morphology typical of Otavi Group limestone and dolomite with associated karst weathering. After the turn-off to Kamanjab, however, the undulating hills are left behind, and the landscape becomes very flat with little or no outcrops. This part of the route is once again underlain by gneisses of the Huab Complex.

After 13 km, a mountain range, which appears on the horizon in the N, indicates the northern boundary of the terrane underlain by Huab Complex lithologies. This mountain range is reached 44 km N of Outjo and consists of limestone and dolomite of the Tsumeb Subgroup. The area between here and the entrance into the Etosha National Park consists of two E-W trending synclines separated and enveloped by anticlines. A synclinal structure which contains Mulden Group phyllites in the core occurs to the N of the road, but unfortunately there are no outcrops, and the phyllites are covered by Cenozoic sediments. Some 23 km N of the first mountain range, a second range of hills, again comprising limestones and dolomites of the Tsumeb Subgroup, separates the two synclines. As the road descends into the second syncline filled with Mulden phyllites and covered by Cenozoic sediments, the northernmost anticline becomes visible on the horizon, particularly in the W. Tsumeb Subgroup limestone and dolomite here form the Ondunduzonananandana Mountains of the southern Etosha National Park.

As the road approaches the entrance to the park, the landscape becomes increasingly flat. The entrance is situated 94 km N of Outjo, and at this point the last outcrops in Tsumeb Group lithologies can be seen. Inside the park, the terrane is underlain by Cenozoic sediments of the Kalahari Group. Okaukuejo, situated on the western fringes of Etosha Pan (see 8.8) is reached 117 km N of Outjo.

## 9.25 Outjo – Kamanjab – Opuwo – Ruacana

by Herbert Roesener and Gabi Schneider

The road from Outjo towards Kamanjab initially crosses an interesting section through folded Otavi and Mulden Group rocks of the Northern Platform of the Damara Orogen, then moves into Palaeo- to Mesoproterozoic rocks, before entering once more terrane underlain by Otavi Group lithologies. It then follows the western edge of the Owambo Basin in an area underlain by Cenozoic sediments of the Kalahari Group, before eventually reaching Ruacana, where Palaeoproterozoic rocks of the Epupa Complex are in contact with metasediments of the Nosib Group of the Damara Sequence (Figs. 9.25.1, 9.25.2).

The area immediately NW of Outjo is occupied by an anticline with older phyllitic schist of the Nosib Group in its core. The next two ridges, 1,5 km and 5 km N of Outjo, consist of Tsumeb Subgroup dolomite which is exposed in road cuts, and forms the limbs of a syncline. The core of the syncline cosists of phyllite of the Mulden Group. The folding in this are is due to NW-SE directed compressional stress that resulted in the formation of the major open folds with NE-trending fold axes, as well as second generation tighter parasitic folds (Thirion 1970).

The Kamanjab Inlier, which starts at the turnoff to Kamanjab, consists mainly of basement rocks of the Huab Complex, which are a succession of metamorphosed sedimentary, volcanic and plutonic rocks. The main part of the journey to Kamanjab shows lit-

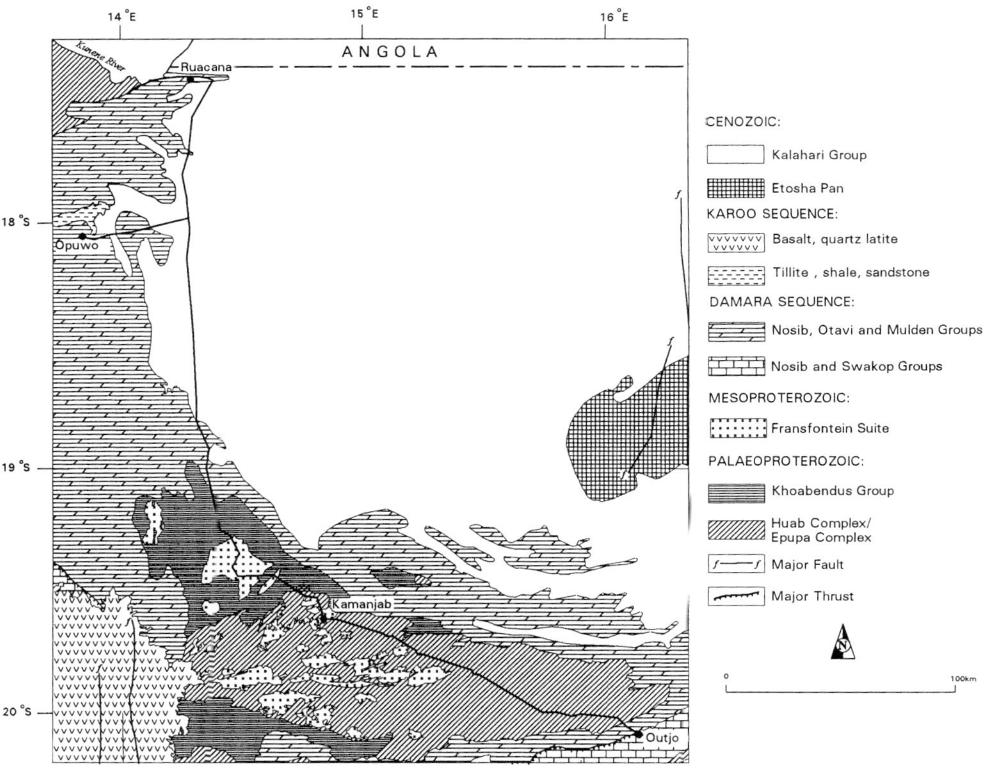


**Fig. 9.25.1:** Geological map for route 9.25.

**Fig. 9.25.2:** Stratigraphic column for route 9.25.

tle outcrop, and it is only closer to Kamanjab that outcrops of granitic and granodioritic rocks of the Huab Complex occur.

The big white road ("Groot Wit Pad"), as it is known, runs N of Kamanjab to Ruacana. The granites and granodiorites of the Huab Complex persist until some 27 km N of Kamanjab, where in a road cut prior to the road going down a slope, rhyolites of the Otjivasandu Member of the Khoabendus Group are exposed. These are interbedded with a white quartzite and folded to form a series of tight anticlines and synclines (Porada 1974). In the lower lying areas, the porhyritic Kaross Granite of the Fransfontein Suite crops out. This granite is red to greenish red and varies from fine to coarse grained.

Some 38 km from Kamanjab the road cuts a small ridge consisting of meta-sediments rimmed by white quartzite of the Khoabendus Group. The road follows this ridge for a couple of kilometers and then crosses back into the Kaross Granite. The road then swings northward and follows the fence of the Etosha Game Park. Most of this stretch is underlain by the Kaross Granite, until the prominent Tevrede ridge is reached 75 km from Kamanjab. This NE-trending ridge rises some 60 m above the surrounding area, and consists of several beds of argillite and white quartzite, some dolomite, limestone and calcareous mudstone forming the central part with again argillite and white quartzite on the northern slopes. A thinly banded cherty iron formation is also present in this area, and the entire sequence belongs to the Khoabendus Group. The white quartzite hosts some gold mineralisation, which is associated with the ferruginous zones.

Further N, the road crosses several different acid volcanic rocks belonging to the Khoabendus Group. It follows an E-W trending ridge consisting of green andesitic tuff overlying white quartzite, until the Werda Veterinary Control Gate is reached. The hills to the W of the gate consist of rhyolite at the base, overlain by the white quartzite of the

Otjivasandu Member. Northwards the road climbs to the top of a plateau where the acid volcanics of the West End Member are overlain by recent sediments. In the Kowares area, the road crosses back into the West End Member igneous rocks, and granites and granodiorites of the Fransfontein Suite form the low hill towards the W.

N of Kowares, some 95 km from Kamanjab, the road traverses the Nosib Group, which is here represented by a red feldspatic quartzite with a basal boulder conglomerate. The road cuts a NW trending ridge of Otavi Group dolomites about 115 km from Kamanjab. The rocks consist from S to N of thinly bedded limestones of the Maieberg Formation, overlain by massive, poorly bedded, light gray dolomites with stromatolitic chert of the Elandshoek Formation, which in turn is overlain by the light grey bedded dolomite of the Hüttenberg Formation. The northern side of this ridge is underlain by reddish-brown, coarse-grained sandstone of the Mulden Group. The next ridge, to the E of the road, is only composed of Elandshoek and Hüttenberg Formation dolomites and is one of several in this area that mark the fold axis of an anticline which can be traced for several tens of kilometers. The hills some 30 km from here and the one close to the Opuwo turn off are expressions of this fold axis.

The road to Opuwo initially passes two mountain ranges composed of Tsumeb Subgroup limestones. It then passes the Joubertberge, also comprising Tsumeb Subgroup lithologies, and descends into dolomites of the Abenab Subgroup. Opuwo is underlain by these dolomites. The prominent flat-topped mountain to the N of Opuwo is a Karoo inselberg composed of tillite, shale and sandstone of the Dwyka Formation.

The journey from the Opuwo turnoff to Ruacana is about 100 km long, and for the most part traverses Kalahari sediments consisting of loose sands and calcretes. The isolated hills scattered in the distance are composed of Otavi dolomites. Closer to the settlement of Ruacana, the Nosib Group sediments form a prominent escarpment. A good section through the Nosib Sediments is exposed along the road to the Ruacana Falls. The upper layers are well bedded and the lower units are composed of a basal conglomerate that contains clasts several tens of centimeters in size. In addition to the Nosib sediments, the younger fluvio-glacial sediments of the Dwyka Formation, Karoo Sequence, crops out in places along the road to the hydro-electric power station.

The Ruacana Falls are situated west of Ruacana at the contact of gneisses of the Epupa Complex and the sediments of the Nosib Group, which are more amenable to erosion. The falls are 700 m wide with a total drop is about 124 m, and the river has cut its way northwards, forming a 120 m deep zig zag gorge, with almost vertical walls. The gorge is some 1.6 km long and follows the contact of the gneiss with the sediments. The valley widens rapidly from a narrow point some 600 m upstream of the falls forming numerous channels inter-fingering with islands of Epupa gneiss. During the rainy season, when the river is in full flood, the falls are continuous, but as discharge decreases significantly during the winter months, the fall becomes a series of individual falls. Downstream from the falls fluvio-glacial sediments of the Dwyka Formation, Karoo Sequence are exposed, indicative that the Kunene River valley was partly shaped by a huge glacier during the Dwyka glaciation some 300 Ma ago. See also Excursion 9.23.

The Calueque Dam, some 12 km upstream of Ruacana, regulates the influx of water to Ruacana, and the diversion weir at Ruacana Falls channels the water to the underground turbines of the 240 MW Ruacana powerstation. This station produces up to 60 % of Namibia's electricity when the Kunene is in flood, but on average it provides 20 % of Namibia's annual power consumption.

## 9.26 Outjo – Khorixas – Uis – Henties Bay

The 367 km long route remains for the larger part in rocks of the Northern Zone of the Damara Sequence. It runs past the spectacular Ugab River terraces and the country's highest mountain, the famous Brandberg, before it crosses into the Central Zone of the Damara Sequence and finally reaches the Namib Desert and the Atlantic coast (Figs. 26.1, 9.26.2).

The town of Outjo is underlain by marble of the Karibib Formation, Swakop Group of the Damara Sequence, while in the area immediately to the NW of the town there is an anticline of phyllitic schist with a core of Chuos iron formation. This anticline forms an E-W ridge, and road C39 towards Khorixas runs immediately to the S. S of the road, there is a wide peneplain underlain by schist of the Kuiseb Formation, and visible in the distance in the SE, is the Paresis Complex, composed of alkaline volcanic and sub-volcanic rocks. Also S of the road, Cenozoic sediments form prominent terraces consisting of conglomerates, cemented by limonitic calcrete, which were deposited by the Ugab River. The Ugab River is one of the larger ephemeral rivers in Namibia and, with its source between Otavi and Outjo, extends for almost 500 km through part of Damaraland, past the Brandberg, and through a spectacular canyon, before reaching the Atlantic Ocean.

From Outjo, for more than 80 km there is little change in the landscape, with the ridge of Karibib Formation marble to the N of the road, and the Ugab River terraces in the S. The occasional iron formation float is the only indication of the Chuos Formation. About 82 km from Outjo, road 2743 turns off to the S and leads to Vingerklip (see 8.31), a prominent erosional remnant in the Ugab River valley well worth the detour of 45 km. As the main road to Khorixas continues westwards, Okonyenya Mountain becomes visible in the distance in the S. The mountain is one of the Cretaceous anorogenic complexes of northwestern Namibia and is composed chiefly of gabbro. Further on, the Brandberg may be visible on a clear day, while close to the road on its northern side a marble ridge consists of rocks of the Tsumeb Subgroup.

The road then turns towards the S about 120 km fom Outjo, and descends slightly where it cuts into a calcrete terrace of the Ugab River, which is exposed in roadcuts. Namibia's highest mountain, the Brandberg (see 8.1), is now the single most prominent feature in the distance in the SW. The junction of roads C35 and C39 is reached some 126 km from Outjo, and while road C39 continues in a westerly direction and reaches

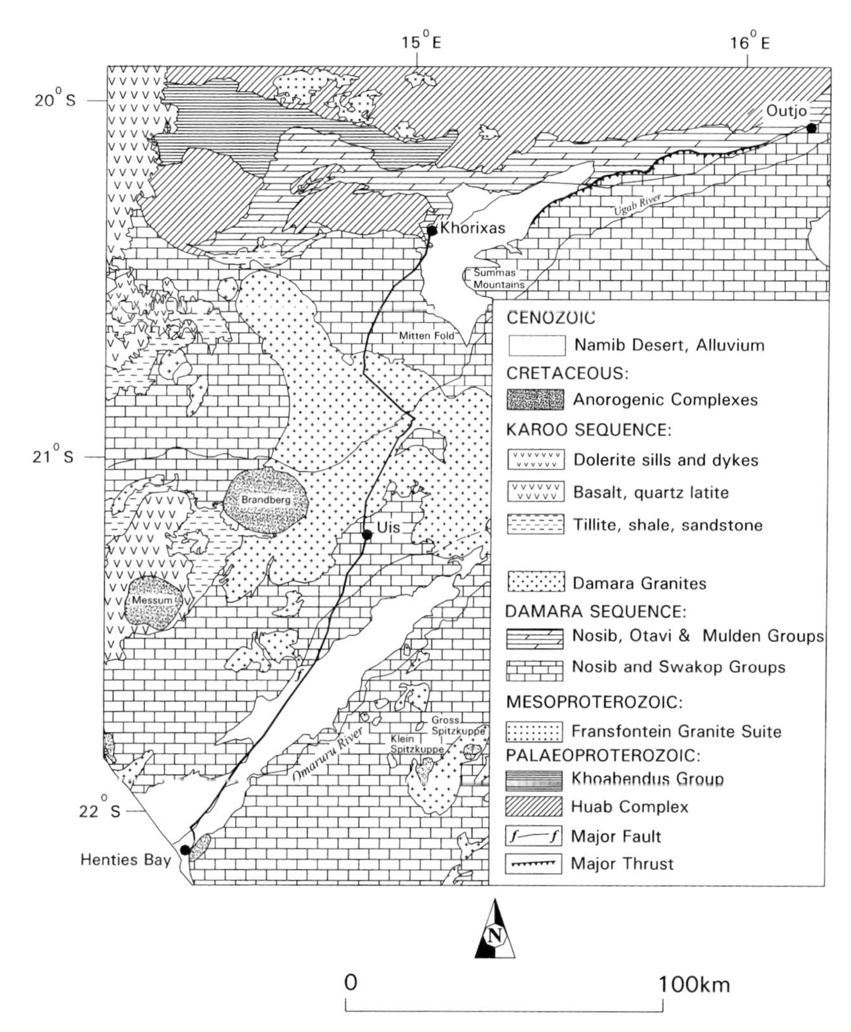


**Fig. 9.26.1:** Geological map for route 9.26.

Khorixas after another 7 km, the route to Uis and Henties Bay follows road C35 in a southerly direction. The ridge seen in the W towards Khorixas is composed early Damaran volcanic rocks and slightly younger Damaran marbles.

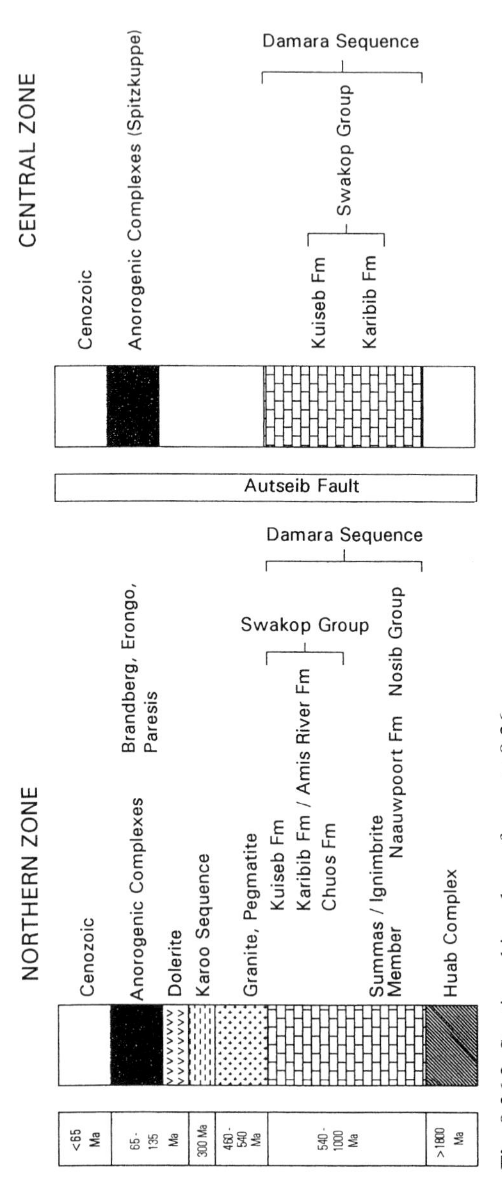


**Fig. 9.26.2:** Stratigraphic column for route 9.26.

After the junction, the road descends rapidly into the upper valley of the Aba-Huab River in an area that is underlain by schist of the Kuiseb Formation, Swakop Group of the Damara Sequence. There is little outcrop, however, as the Neoproterozoic rocks are concealed by the Cenozoic sediments of the river. The river is crossed some 28 km from the junction, and as the road climbs out of the valley, the hills on both sides are composed of Damaran Salem-type granite. This granite forms the country rock of the larger part of the remaining route to Uis, and is a syn- to post-tectonic, coarsely porphyritic biotite-granite. The higher mountains in the distance in the E are the Summas Mountains and the Mitten Fold. They consist of alkaline volcanics of the Naauwpoort Formation, Nosib Group, which were deposited in a northern graben during the early stages of rifting. The Summas Ignimbrite Member extruded along the Summas Fault, and is made up of a lower, over 6000 m thick succession of ignimbrite and minor quartz-feldspar porphyry lava, with an upper portion containing several layers of vesicular basalt. Still further E, the Okonyenya Mountain is once again visible.

Some 50 km from the junction, the road makes a marked turn towards the SE, and descends into the valley of the Ugab River. While the landscape is completely dominated by the Brandberg in the SW, the Ugab River terraces are well exposed ahead of the road (see 8.31), and the view to the E contains the Mitten Fold and Okonyenya Mountain, which, at 1902 m above sea level, is by far the highest peak in the E. S of Okonyenya, the Otjongundu Plateau in the Otjihorongo area is composed of mudstone of the Omingonde Formation, Karoo Sequence, topped by harder trachyte sills, which protected the mudstone from weathering.

The Ugab River is crossed by a bridge about 68 km from the juction, and after the bridge, the route resumes a southwesterly course. More hills, seen in the E, are composed of Naauwpoort volcanics, and the Ugab River valley forms a prominent depression in the NW. About 20 km from the bridge, the Uis Mountains, composed of Cretaceous dolerite intruded into Salem granite, appear in the S. After another 8 km, road 2359 turns off to the W and leads to the Amis Valley in the Brandberg (see 8.1), which hosts the famous “White Lady” rock art.

Meta-greywackes of the Amis River Formation, Swakop Group of the Damara Sequence are exposed at Uis, which is reached 119 km from the junction E of Khorixas. Pegmatite swarms, containing cassiterite and tantalite mineralisation within greisen zones, cut the meta- greywackes. These were mined at Uis from 1911 until 1988 when the mine closed due to the collapse of the tin price and the high operational costs related to the low grade of the ore. The Uis Tin Mine, when operational, was the world’s largest hard rock tin mine, and it still has ore reserves of 72 million t grading 0,1 – 0,15 % tin.

From Uis, the road heads in a southwesterly direction towards Henties Bay on the Atlantic coast, and just after the mine, apart from the towering Brandberg, the landscape is undulating and underlain by meta-greywackes that have been intensely intruded by granites and pegmatites exposed in hills. Towards the E, the higher mountains are composed of marbles of the Karibib Formation, Swakop Group. These mountains represent outcrops of Central Zone rocks, as they occur E of the Autseib Fault, which separates the Central and Northern Zones of the Damara Orogen in the area. About 9 km from Uis, the turn-off to the now defunct Brandberg West Mine is reached. Tin- and tungsten-bearing quartz veins in sediments of the Ugab Subgroup, Damara Sequence, were mined here between 1946 and 1980 in an open cast operation. The ore mineral assemblage includes cassiterite, wolframite, scheelite, chalcopyrite, sphalerite, pyrrhotite, galena and marcasite.

Although there is no related expression in the landscape, the Autseib Fault is crossed some 13 km from the turn-off. The road now follows a flat peneplain underlain by schist of the Kuiseb Formation and, occasionally, marble of the Karibib Formation. It soon crosses an area underlain by Cenozoic sediments deposited by the Omaruru River, and the landscape, apart from the ever-present Brandberg in the N, becomes even more featureless. The Omaruru River, with its catchment area around Kalkfeld and Omaruru, is one of the large ephemeral rivers that drains into the Atlantic Ocean. The catchment area usually receives good rainfall, and the Omaruru River is therefore often one of the first rivers to flow during good rainy seasons. A dam scheme was built in the 1980s some 35 km upstream from the mouth of the Omaruru River, where the floods of the river are temporarily stored. Once the fine sediment in the dam water has settled, the water is allowed to flow slowly out of the dam into the coarse sand of the riverbed below the wall. This quickly disappears into the sand and flows underground, where it is protected from evaporation, to feed the aquifers of the Omaruru River delta. Numerous boreholes tap these aquifers for the water supplies of Henties Bay and Swakopmund.

The Grosse and Kleine Spitzkuppe (see 8.29) become visible in the distance in the SE after some 50 km and some 102 km from Uis road C35 joins the coastal C34 road. Turning S, the town of Henties Bay is reached after some 7 km, just after crossing the Omaruru River. The town itself is located in an area underlain by a southern channel of the extensive Omaruru River delta.

## 9.27 Swakopmund – Henties Bay – Cape Cross – Terrace Bay

The route from Swakopmund to Terrace Bay runs for 347 km along the Namibian Atlantic coast, and gives a good view of the various landscapes found in the Cenozoic of the Namib Desert. The underlying country rocks cover a geological period from the Late Proterozoic to the Early Cretaceous, and while the road mainly follows terrane underlain by Cenozoic surficial sediments, rocky outcrops along the shoreline have formed a number of bays which can be seen along the route (Figs.9.27.1, 9.27.2).

Just N of Swakopmund, the road follows the coastal peneplain, which is underlain by marbles of the Swakop Group. Some 5 km out of town, the Swakopmund Salt Works become visible on the western side of the road. These salt works make use of some of the abundant salt pans that exist along the Namibian coast, which are floored by a seal-

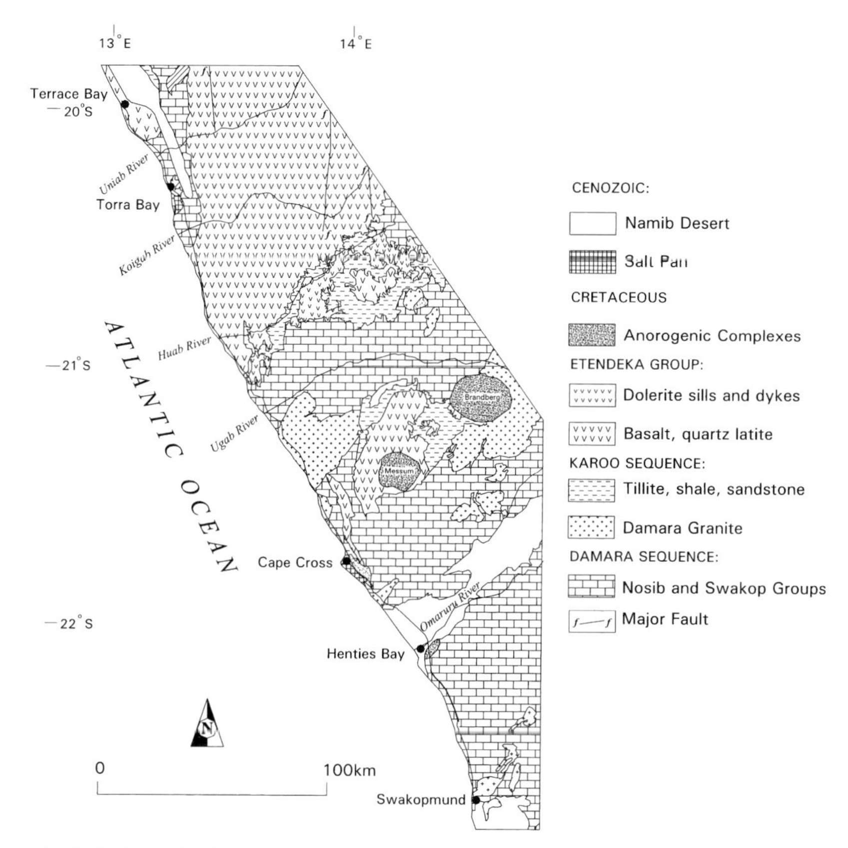


**Fig. 9.27.1:** Geological map for route 9.27.

ing layer of gypsum and clay. Seawater is pumped into these pans, and due to the high rate of evaporation in the arid climate, coupled with an almost complete lack of rainfall, the water evaporates from the pans leaving the salt. It is then harvested with bulldozers and refined. The Swakopmund Salt Works started operation in 1936 and have an annual production of some 150 000 t of salt. Some of this salt is refined to produce iodised table salt. Utilising the abundance of sea birds, a guano platform has been erected in one of the evaporation ponds, and produces about 1 000 t of the phosphate rich fertiliser per annum.

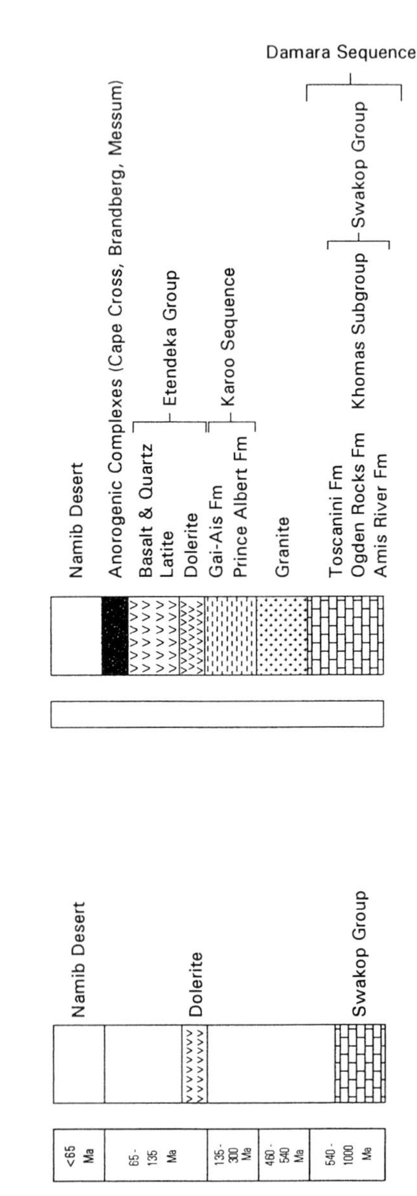


**Fig. 9.27.2:** Stratigraphic column for route 9.27.

Some 13 km after the salt works, a gypsum pan can be recognised on the eastern side of the road. The ground has a distinct darker colour than the surroundings, and a fragmented texture. Gypsum is formed in abundance along the Namibian coast through a reaction of sulphur contained in the fog and calcium carbonate derived from Damaran marbles or calcrete. The sulfur in the fog is derived from the frequent "sulphur eruptions" caused by the periodic mass mortality of plancton in the Atlantic Ocean (Martin 1963). The road continues northwards in Cenozoic sediments of the Namib Desert, and the landscape is only broken by the occasional dolerite dyke. These dolerites are of Early Cretaceous age and have developed through extensional tectonics associated with the opening of the South Atlantic Ocean during the break-up of Gondwanaland.

The settlement of Wlotzkasbaken is reached 30 km N of Swakopmund. The small bay and the rocky outcrops to the E of the road are composed of a coarse variety of dolerite, the so-called "trout-stone" or troctolite, which has been tested for quarrying ornamental stone, but so far no production has occurred. Just after Wlotskasbaken the road passes through one of the most extensive lichen fields along the entire coast. This symbiosis of an algae and a fungus grows on the surface of the rubble strewn around, and thrives solely on the moisture contained in the coastal fog.

The coastal town of Henties Bay is reached some 67 km N of Swakopmund, and on a clear day the Brandberg (see 8.1) is visible to the NE. The town itself has been built in an area underlain by a southern channel of the extensive Omaruru River delta. The Omaruru River is one of the large ephemeral rivers that drains into the Atlantic Ocean, with its catchment area around Kalkfeld and Omaruru. This area usually receives good rainfall, and the Omaruru River is therefore often one of the first rivers to flow during good rainy seasons. A dam scheme was built in the 1980s some 35 km upstream from the Omaruru Delta, where the floods of the river are temporarily stored. Once the fine sediment in the dam water has settled, the water is allowed to flow slowly out of the dam into the coarse sand of the river bed below the wall. This quickly disappears into the sand and flows underground, where it is protected from evaporation, to feed the aquifers of the Omaruru River delta. Numerous boreholes tap these aquifers for the water supplies of Henties Bay and Swakopmund. On leaving Henties Bay the road crosses the river just outside the town.

Further N, the scenary remains quite monotonous in a peneplain underlain by surficial deposits which also conceal the western extension of the Autseib Fault, which marks the transition between the Central Zone and the Northern Zone of the Damara Orogen. The Cape Cross Salt Pan is reached 35 km N of Henties Bay and numerous dolerite dykes together with the southernmost volcanics of the Etendeka Formation (see 8.6) are encountered next to the road.

As the road continues, the large Cape Cross Pan extends between the road and the sea and some 52 km N of Henties Bay, in an area where the pan comes fairly close to the road, infiltrating seawater is evaporated in a small artificial excavation, close to the road, providing a good view of the crystallisation process. At this locality, the hills seen in the E are composed of gabbroic rocks of the Cape Cross Complex and Etendeka basalts.

The Cape Cross Salt Works are reached 65 km N of Henties Bay. Here, fossil salt that has formed in an ancient lagoon is extracted, and production is supported by man-made excavations into which seawater is captured and allowed to evaporate forming salt. Total production is about 6 000 t per annum. Some 4 km further the road passes the turn-off to Cape Cross. This prominent cape is formed by granite of the Cape Cross Complex, the westernmost of the Cretaceous anorogenic igneous complexes that extend in a line from Cape Cross to N of Otjiwarongo. It has a complex lithology, comprising nepheline syenite, syenite, granite and gabbro. The very rare boron silicate jerejemevite is found here, and the only other occurrences worldwide are at Erongo and in an alkaline complex on the Kola Peninsula. The rocks of the Cape Cross Complex are also home to the largest known colony of the Cape Fur Seal *(Pusilis pusilis)*. N of Cape Cross, a sheltered bay has been used as a harbour in the past, which supported guano harvesting, as well as sealing.

As the road continues northwards, the Albin Ridge runs parallel to the coast in the E. In this ridge, sediments of the Prince Albert and Gai-As Formations of the Karoo Sequence are unconformably overlain by Etendeka Formation volcanics. The Karoo sediments comprise bands and lenses of sandstone and a conglomerate with poorly sorted clasts of sandstone, basalt, quartz and schist. The volcanics have a thickness of 200 m consisting of basalt capped by quartz latite. The entire succession is intruded by the Horing Bay dolerites, a suite of dolerite dykes and sills, typically less than 1 m in thickness. The Albin Ridge provides a section through rocks that were deposited during the early rifting stages leading to the formation of the Atlantic Ocean during the break-up of Gondwanaland. Such early rifting is characterised by volcanic activity and the deposition of sediments that have been transported over small distances and are therefore poorly sorted. The conglomerates which immediately underly the basalts are syn-volcanic and represent the earliest rift deposits (Milner & Swart 1994).

Approaching Horing Bay some 22 km N of Cape Cross, the road enters an area underlain by coarse grained-biotite granite of Damaran age. Schistose turbidites of the Amis River Formation, Swakop Group of the Damara Sequence crop out in the fairly flat area to the NE of the granite, while the Albin Ridge can still be seen to the E. Dolerite rubble from the Albin Ridge covers the flat surfaces on both sides of the road. Road C2303 to the old Brandberg West tin-tungsten mine turns off to the E, some 30 km N of Cape Cross, and it also provides access to the Messum Complex (see 8.18). The Mile 108 campsite is reached some 41 km N of Cape Cross. A marine pebble terrace, which is extensively cemented by gypsum in places, underlies the area, and the pebbles are representative of the rocks located in the E.

As the road continues N of Mile 108, the landscape is dominated by prominent dolerite dykes running parallel to the coast seen on either side of the road. Some isolated ridges of Damaran granite also occur, and the Brandberg (see 8.1) is still visible in the E on clear days. Interestingly, the sparse dwarf shrubs, comprising mostly of the salt bush *Salsola*, gather sand on their southwestern sides, due to the prevailing winds. Some 16 km N of Mile 108 there is a distinct change in colour of the rubble, which is now comprised of Damaran leucogranite. The extensive Ugab Salt Pan, which in the past supported the Ugab Salt Works, occurs between the road and the sea. Dolerite dykes continue to be prominent on both sides of the road, and about 30 km N of Mile 108 the turbidites of the Zerissene Mountains become visible in the distance in the E.

The entrance to the Skeleton Coast Park at Ugabmund is reached 35 km N of Mile 108. The area here is underlain by grey, metamorphosed, siliceous turbidites of the Amis River Formation, and there are extensive outcrops on both banks of the Ugab River between the road and the sea. The Ugab River is crossed immediately after the gate. The river bed of this ephemeral stream is more than a kilometer wide and is marked by an abundance of vegetation, mostly tamarisks. N of the river the landscape is once again dominated by dolerite ridges, with the occasional outcrop of the turbidites. Some 6 km N of the Skeleton Coast Park gate, a marine pebble terrace forms a prominent platform to the NW of the road.

Some 2 km further on, the dark dolerite ridges stand out in stark contrast to the light coloured desert surface of the marine terrace in the SW. The road then approaches the mylonites of the Ogden Rocks Formation, Khomas Subgroup, Swakop Group of the Damara Orogen. Blastomylonites and marbles form the ridges on either side of the road in a zone of intensive shearing and thrusting. The shipwreck of the South West Seal lies between the sea and the western ridge, and can be viewed from a lookout point some 16 km N of Ugabmund, and after another 4 km, a prominent outcrop of dolerite appears in the E.

The road then continues in an area underlain by basalt of the Etendeka Formation, and about 42 km N of Ugabmund, an abandoned drill rig for drilling an exploration borehole for oil near the mouth of the Huab River is encountered. The surrounding extremely flat area is underlain by Cenozoic surficial sediments, and the surface is strewn with basalt rubble. The hole was drilled in 1972, and intersected some 760 m of Karoo sediments, including some coal. It was drilled to a total depth of 1735 m, where it stopped in rocks of the Damara Sequence. It is no surprise that oil was not discovered, and today the old equipment is a breeding area for the Cape Cormorant. Some 5 km further on, a large salt pan becomes visible in the E. Between the road and the sea some large trenches bear witness to diamond exploration that took place on the marine terraces, W of the road. Recently, this exploration has been revived, using modern earth moving machinery.

About 52 km N of Ugabmund a prominent ridge to the E of the road is composed of Etendeka Formation volcanics in contact with a greenish chlorite-muscovite schist of the Toscanini Formation of the Swakop Group, Damara Sequence. The volcanics also form the ridge seen in the W. Between the ridges on either side of the road, there are some small salt pans, one of which is larger than the rest. Further W of the western volcanic ridge can be seen the remains of diamond mining activities on the marine terrace, where in 1962/3 some 400 carats of diamonds were recovered. In 1972, a new recovery plant was erected, however, for unknown reasons not a single diamond was recovered.

Coal is also known to occur in the Karoo rocks of this area, although so far limited exploration has not been able to detect economic quantities.

Some 4 km further on, ridges of Etendeka volcanic rocks with sand on their western flanks still dominate the landscape on either side of the road. Further on, they form the eastern margin of a marine terrace, and erosional channels within the volcanics are clearly marked with accumulations of light coloured sand. To the W of the road Etendeka Formation volcanics form ridges high enough to partially block the sea view.

Some 16 km N of Toscanini, the landscape is dominated by rolling hills of Etendeka Formation basalts on either side of the road. A prominent ridge of Etendeka Formation quartz latites (see 8.6) occurs to the E, and the typical weathering pattern of volcanic rocks can be observed. The distant mountain range in the E is also composed of Etendeka Formation basalts. After another 2 km, Etendeka Formation quartz latite also forms a ridge between the road and the sea, and the land surface is strewn with rubble from these volcanic rocks.

A distinctly different dark outcrop seen W of the road some 32 km N of Toscanini is composed of migmatised greywackes of the Khomas Subgroup, Swakop Group of the Damara Sequence. These rocks show a typical circular weathering pattern which can also be observed further N at Palgrave Point. The road itself now follows a peneplain underlain by Cenozoic surficial sediments of the Koigab River delta. The mountains seen in the E are part of the main Etendeka lava field. The turn-off to Koigabmund is passed about 44 km N of Toscanini, where the meta-greywackes mentioned above now form a distinct ridge N of the road. The road then ascends onto an area underlain by the meta-greywackes, where ridges form a rather rugged terrane. Thereafter the road descends once more onto a peneplain underlain by Cenozoic surficial sediments, with abundat outcrops of meta-greywacke to the E of the road. High, coast-parallel sand dunes can be seen in the N.

The turn-off C39 to Springbokwater and Khorixas is passed 53 km N of Toscanini. The area is characterised by numerous, very mobile barchan dunes, which often cross the road, while in the distance large longitudinal dunes can be seen in the E. These longitudinal dunes have black and reddish heavy mineral concentrations on their western flanks, indicating the presence of ilmenite, magnetite and garnet within the sands. There are still some outcrops of meta-greywacke closer to the road, and as always the ever present dolerite dykes. One particularly spectacular dyke runs parallel to the road just after the Springbokwater turn-off. After the turn-off there are a few more outcrops of the meta-graywacke on the eastern side of the road, while to the W there are the first occurrences of the Torra Bay granite, a coarse-grained hornblende-biotite-garnet granite of Damaran age. Approaching Torra Bay, the granite outcrops become more abundant, and after a marked turn to the W, the road enters terrane underlain by granite. In outcrop the granite has a migmatitic appearance with abundant light-coloured veins and dark inclusions. Torra Bay is reached about 10 km N of the turn-off to Springbokwater.

N of Torra Bay, the road once more enters a peneplain underlain by Cenozoic sediments, interrupted by the occasional dolerite dyke. The surface is strewn with rubble from the Etendeka volcanics to the E, where the large longitudinal dune still runs coast-parallel, and some 10 km N of Torra Bay the extensive delta of the Uniab River is reached, and five main arms of this river are crossed. Some of these channels have permanent water, which together with the abundant vegetation attracts game, such as oryx, springbok and ostrich for example. Smaller hills seen on both sides of the road are composed of meta-greywacke, and the turn-off to Uniabmund, where the main channel of the Uniab River enters the sea, is 15 km N of Torra Bay.

The Uniab River, with its headwaters in Damaraland, once formed a large delta, which, now well raised above sea level, has been cut by the flowing river, forming five main watercourses. Today, only one of the channels is used, although in the others water does seep underground forming a number of pools surrounded by reeds and other vegetation. When rain upstream is sufficient, the Uniab flows in the second most northern channel, and it is from the bed of this part of the Uniab that water is extracted for use at Terrace Bay and Torra Bay. There is also a waterfall in this channel. The trickle of water, which is usually present on the surface of the river bed, drips and splashes down from the grey desert surface through a variety of attractive red sandstones and yellow calcrete layers. It continues through a narrow canyon until the water seeps into the sands and pebbles of the beach. This waterfall alone is more than just a reward for the half-hour walk from the road (Fig. 9.27.3).

**Fig. 9.27.3:** The waterfall at Uniabmund.

After crossing of the main channel of the Uniab River, the road continues for another 4 km to cross other smaller channels of this major delta system. Vegetation remains lush, and even the dunes to the E of the road are vegetated. W of the road the incised river beds provide ample outcrop of the Cenozoic sediments deposited by the Uniab River, and for the next 5 km the road remains in a peneplain underlain by Damaran rocks. Then a change in morphology indicates that the coastal plain is now underlain by Etendeka volcanics, and there are some small, isolated volcanic hills. The coastal dune belt remains prominent in the E, and some 14 km N of the Uniab River, the road crosses a ridge of Etendeka volcanics. It then runs very close to the beach, where a pebble terrace can be observed. The vast peneplain E of the road is underlain by volcanics.

As the road moves inland once more, it now runs between two ridges of volcanics, and mega-trenching from previous diamond exploration can be seen on the western ridge. Further on, salt pans appear on the eastern side of the road, and the western ridge is broken at Deka Bay to provide a view of the sea. As Terrace Bay is approached, the terrane becomes more pronounced with hills of Etendeka volcanics, obscuring both the sea view and the coastal dune belt. Terrace Bay is reached 32 km N of the Uniab River. It is noteworthy that the largest hill at Terrace Bay is man-made consisting of tailings from previous diamond production. Diamond mining took place here from 1957 to 1967, and some 10 000 carats of diamonds were produced. Later, between 1970 and 1971, a dense media separation plant erected at Terrace Bay recovered some 2 500 carats from the old tailings and from new terrace material. The plant has since been dismantled and only the mine dump remains, although some of the buildings today used by Nature Conservation date back to the diamond mining activities.

## 9.28 Khorixas – Palmwag – Sesfontein

This 277 km long route starts in Neoproterozoic Damaran rocks, passes a section through the Etendeka volcanics of the Karoo Sequence and ends in terrane underlain again by Damaran rocks (Figs. 9.28.1, 9.28.2). As the road leaves Khorixas westwards, the mountain range to the N of the road is composed of granite of the Palaeoproterozoic Huab Complex. To the S, the terrane is underlain by schist and marble of the Swakop Group, and the dome-shaped hills in the background are formed by Damaran granite. Some 5 km outside Khorixas, the valley widens, and undulating hills of Damaran schist appear on both sides of the road. The road itself follows the course of the Aba-Huab River in an area that is underlain by recent calcrete, although outcrops of the bedrock geology occur where the river has incised its bed into the schist.

A completely different type of landscape appears with the first Karoo Sequence rocks visible in the distance, about 16 km from Khorixas to the NW. These typical table-topped mountains consist of flat-lying sandstone and shale layers, with the sandstone being more resistant to weathering causing the table-top mountain topography. The un-

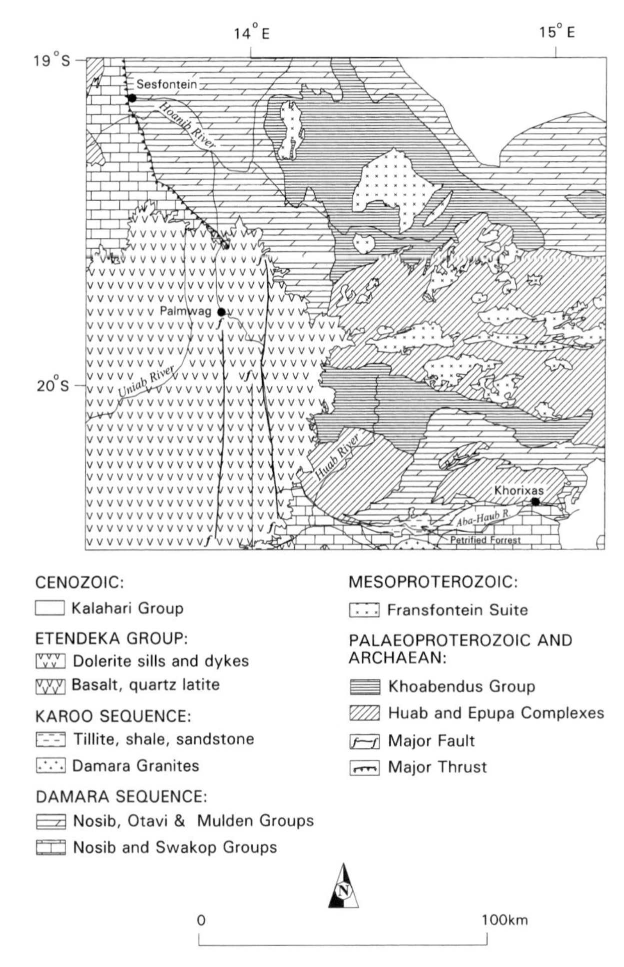


**Fig. 9.28.1:** Geological map for route 9.28.

dulating hills on both sides of the road are still composed of Damaran schist with Huab gneiss in the distance in the N. The road continues in the valley of the Aba-Huab River, and moving westwards, the landscape becomes more and more dominated by the flat-topped Karoo mountains.

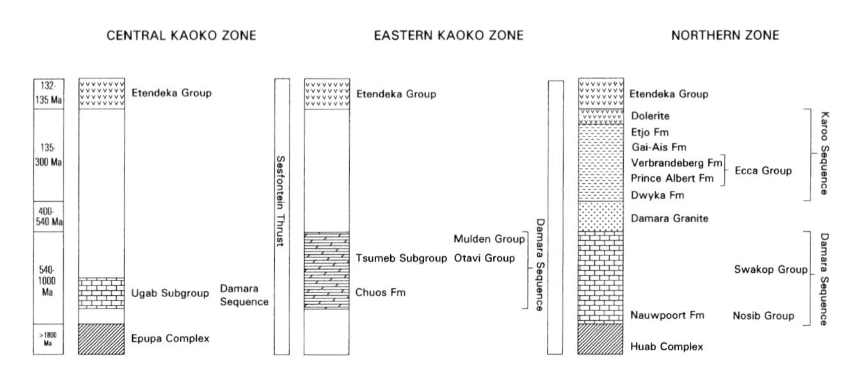


**Fig. 9.28.2:** Stratigraphic column for route 9.28.

A small hill composed of Karoo sediments is visible where road 2328 turns off to the S some 31 km W of Khorixas. On a clear day, the Brandberg (see 8.1) is visible from here in the S, and as the road continues, the hills on the southern flanks of the valley are composed of Karoo sediments overlying Damaran schist. The contact is clearly visible and represents a hiatus of some 300 Ma!

The rocks forming the table-topped montains belong to the Permian to Jurassic Karoo Sequence that started with the deposition of glacial sediments of the Dwyka Formation about 300 Ma ago. The climate warmed over the next 55 Ma and the area was covered by dense vegetation along rivers that formed deltas along the coast of an ocean opening westwards. The sediments deposited at that time are now seen as interbedded shales, siltstones and sandstones of the Prince Albert, Verbrandeberg and Gai-As Formations. Plant remains are visible at the Petrified Forest, located 43 km W of Khorixas (see 8.26), where uprooted tree trunks were deposited in an embayment along a river course and also around Burnt Mountain (see 8.3), where thin coaly layers are interbedded with mudstone and shale.

During the Triassic, the climate had warmed to such an extent, that huge areas of what is now southern Africa and then adjoining South America were covered by a desert with sand dunes. The resulting aeolian sandstone of the Etjo Formation of the upper Karoo Sequence is therefore characterized by prominent large-scale dune cross-bedding. The westerly palaeo-wind that deposited the Etjo sandstone originated from a high pressure cell to the W, and penetrated deeply into Africa, due to the absence of mountain barriers (Bigella 1970). Towards the end of the Karoo period, about 135 Ma ago, dolerite intruded the black carbonaceous rocks of the Permian Prince Albert Formation.

The Aba-Huab valley, in the vicinity of the Petrified Forest, is underlain by Karoo sediments, and three table mountains dominate the landscape. To the N, however, the Naauwpoort volcanics of the Nosib Group, Damara Sequence, now form the hills, and, likewise, the hills to the S of the road, are of Damaran schist with dome-shaped Damaran granite in the distance. After the Petrified Forest, the road passes between Karoo mountains, with the Ecca sediments, containing massive, white, cliff-forming sandstones at the bottom, overlain by massive Etjo sandstone forming the prominent cliffs of the upper third of the mountains. The Etjo sandstone in turn is overlain by a layer of Etendeka volcanic rocks. The road then continues in terrane underlain by Karoo sediments, with Damaran schist forming the undulating hills to the S.

Some 7 km from the Petrified Forest, the road climbs to reach a small pass, from where the main Etendeka lava field (see 8.6) can be viewed. To the S of the road, white Ecca sediments are overlain by the typically red Etjo sandstone, producing the abundant rubble strewn over the landscape which is now dominated entirely by the Karoo Sequence. Further on, the contact between the Damaran basement, which dips at an angle to the SW and the overlying horizontal Karoo sediments is clearly seen in an isolated mountain to the NW of the road.

Some 10 km after the Petrified Forest, the Etjo sandstone forms a number of isolated fossil dunes with well developed aeolian cross bedding. Good outcrops of Etjo sandstone occur close to the road, whilst ahead, the hills have a different morphology from the surrounding Karoo rocks, and are composed of schist of the Damara Sequence. About 20 km W of the Petrified Forest, the mountain range parallel to the road in the S is composed of schists of the Swakop Group, Damara Sequence, while the hills on the northern side of the road belong to the Huab Complex. The turn-off to Twyfelfontein (see 8.30), Burnt Mountain (8.3) and the Organ Pipes (8.24) follows after another 10 km and the road now approaches the main Etendeka lava field, which dominates the view to the W. After the Twyfelfontein turn-off, the road descends into the valley of the Huab, with the rocks on either side comprising Damaran and Huab Complex lithologies. The accumulation of sand dunes around plants, and on the flanks of the mountains is noteworthy. The Huab River is crossed about 20 km from the Twyfelfontein turn-off, and W of the Huab, the valley narrows, and Damaran schist crops out on both sides of the road.

Some 7 km beyond the Huab River, a relic dune of Etjo sandstone with well developed aeolian cross bedding occurs in the middle of the valley on the northern side of the road. The Etendeka volcanics now form the fairly high mountains on both sides of the road, and the *Euphorbia* bushes, which typically thrive on the volcanic rocks, are abundant. The road now climbs an extremely steep pass, the top of which is reached 13 km after the Huab River crossing. Road cuts along the ascent expose Damaran schist at the bottom, and, after passing the contact between the these Neoproterozoic rocks and the Early Cretaceous volcanics, Etendeka basalts and quartz latites. The well developed layering in the volcanics represents individual lava flows.

The road then continues for the next 100 km in Etendeka volcanics, and Palmwag is reached some 162 km from Khorixas. The road negotiates a small pass some 32 km N of Palmwag, and the Damaran rocks to the N of the main Etendeka lava field become visible for the first time. The northern limit of the volcanics is reached about 41 km N of Palmwag, and the road continues for a short distance in terrane underlain by Chuos

Formation mixtites, forming gently undulating hills with little or no outcrops. The prominent mountain range in the NE is composed of Otavi Group carbonates, with Mulden Group sediments along their flanks within synclines plunging towards the S. The contact between these two lithologies is well marked by a change in vegetation. Meanwhile to the W, Etendeka volcanics are still prominent.

About 48 km N of Palmwag, the road descends into a wide valley which contains the major Sesfontein Thrust, along which rocks of the Chuos Formation are thrust onto the Mulden Group. The road proceeds along the eastern side of the valley which is floored by sediments of the Mulden Group, which can have a diverse sequence of rock types varying from very coarse conglomerates to quartzites, phyllites and marl. In this locality, the Mulden Group consists mainly of phyllite and quartzite. It is probable that this formation represents an intermontane basin of limited extent which developed syn-tectonically.The Mulden unconformably overlies the rocks of the Otavi Group. The carbonates of the Otavi Group form an impressive mountain range to the E, with the contact to the Mulden Group well marked by the virtual disappearance of vegetation on the Otavi Group rocks. W of the road, the Mulden forms small hills.

The road to Khowarib Schlucht, where the Hoanib has incised its bed into the mountain range to the E of the road, turns off to the E about 76 km N of Palmwag. Outcrops of Mulden sediments are found right next to the road, and Khowarib Schlucht itself provides excellent exposures of the dolomites of the Otavi Group. A thick package of white, pale calcareous or dolomitic micrite is overlain by the thinly bedded, gray, cherty dolomitic micrite. Interlayered are stromatolitic dolomites and dolomitic sedimentary breccias. These lithologies are interpreted to represent a platform facies at the margin of the Congo Craton. Thrusting, observed in the western Khowarib Schlucht, is syntectonic and much of it has been subsequently folded and overthrust. The thrusting represents crustal shortening from the W, and thrust soles can be identified in the section. At one location talc has formed indicating significant heat generation during the process.

About 1 km from Khowarib Schlucht, the road crosses the Hoanib River, and after another 10 km reaches Warmquelle, where a small settlement is supported by spring water. The spring occurs at the contact of the Otavi and Mulden Groups, and irrigation of crops has taken place since the German colonial period. From Warmquelle, the road follows a valley floored by Mulden sediments, and approaches Gamgurib Mountain, which is formed by an anticline of Tsumeb Subgroup rocks, and once again, substantial folding and thrusting can be observed (Fig. 9.28.3). The Hoanib River cuts into these rocks, which testifies to the substantial age of this drainage system. It is possible that this system drained Lake Etosha through much of the Kalahari times.

Having left the gorge at Gamgurib Mountain, the road again crosses the Hoanib River, which here has a deeply incised river bed with plenty of large trees. It then follows a peneplain once again underlain by Mulden sediments, before Sesfontein is reached 108 km N of Palmwag. Sesfontein is situated on the western fringes of the peneplain, just below the Sesfontein thrust, where the older carbonate rocks of the Tsumeb Subgroup have been thrust onto the younger Mulden sediments. Along the thrust there are

**Fig. 9.28.3:** Gamgurib Mountain.

six springs, which have given the place its name. The availability of water supported settlements here for some time and a fort was built during the German colonial period. The German settlers also implemented an extensive irrigation scheme, and the old channels, as well as the fig trees and imported date palms, which have grown to a substantial height, survive to this day.

The Sesfontein Thrust, which is split up into an upper and a lower thrust, can be seen in the mountains to the W of Sesfontein. Along the upper thrust, sediments of the Ugab Subgroup are thrust onto Tsumeb Subgroup lithologies, while along the lower thrust, the Tsumeb Subgroup is thrust onto the Mulden rocks. The high mountains seen in the NW are composed of gneisses of the Epupa Complex.

## 9.29 Sesfontein – Opuwo – Epupa – Marienfluss – Sesfontein

by Gabi Schneider and Bernd Teigler

This 870 km long circular route takes one to the extreme northwestern parts of Namibia, which are only accessible by specialized vehicles. It begins in rocks of the Neoproterozoic Damara Sequence in the Eastern Kaoko Zone, traverses the oldest crustal domains in Namibia, the Epupa and Kunene Complexes, and, before it returns to Dama-

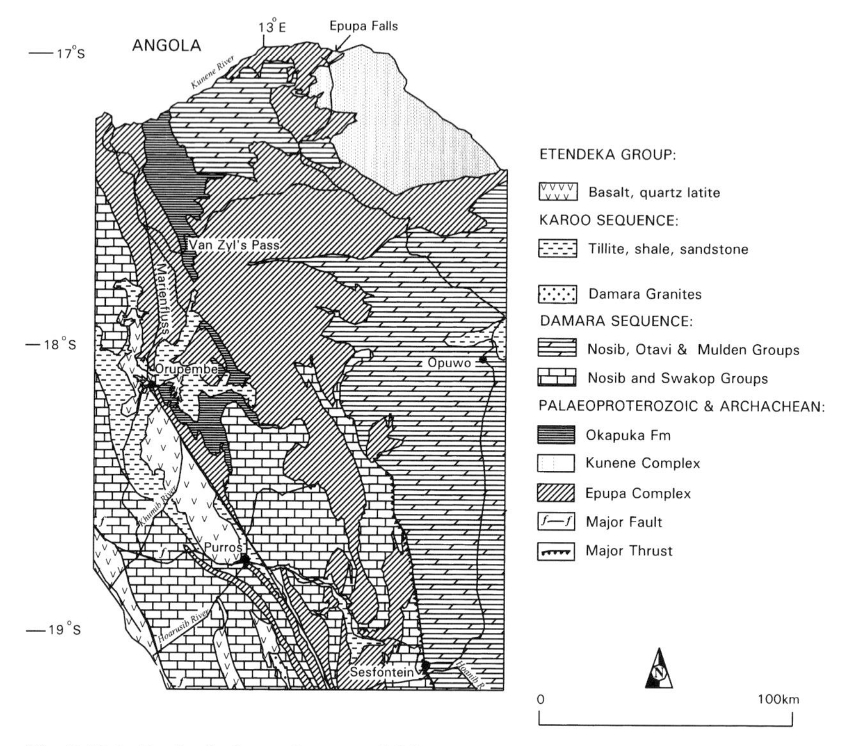


**Fig. 2.29.1:** Geological map for route 9.29.

ran terrane in the Western Kaoko Zone, runs past Karoo Sequence sediments and the volcanics of the Etendeka Group. Before it returns to Sesfontein, it traverses rocks of the Central Kaoko Zone of the Damara Orogen.

The roads are all un-surfaced, and, in contrast to the other routes described in this book, in parts are mere tracks over very rough terrane. A 4 x 4 vehicle is essential, combined with advanced off-road driving skills. To respect the extremely sensitive nature of the arid environment, driving off the track is strictly prohibited. There are also no fueling stations outside Sesfontein and Opuwo, and petrol, water, food and all other supplies, including a minimum of two spare wheels, are essential and have to be carried.

Leaving Sesfontein towards the E along road D3706, the landscape is underlain by sediments of the Mulden Group and Cenozoic sediments deposited by the Hoanib River, which follows the road to the S. After some 12 km, the road approaches the gorge at the

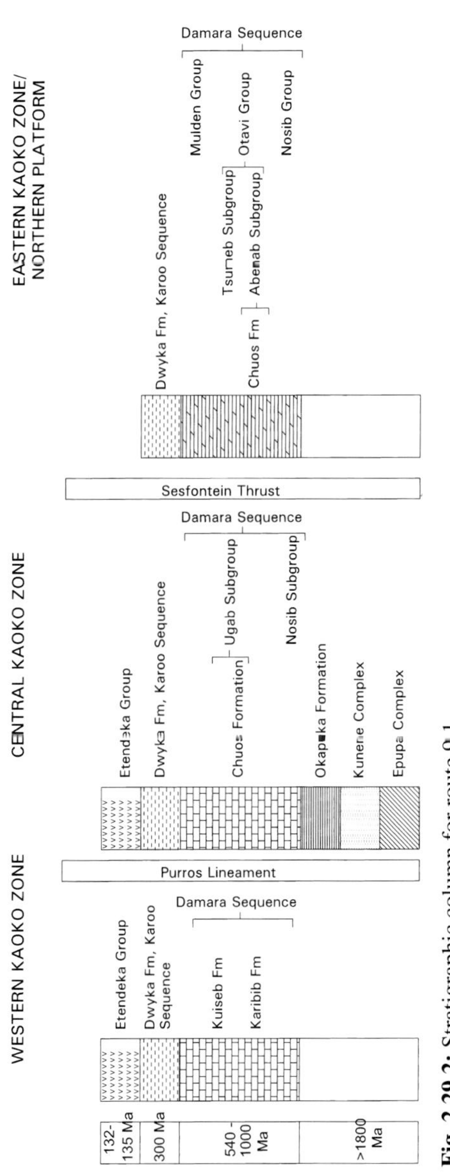


**Fig. 2.29.2:** Stratigraphic column for route 9.1.

Gamgurib Mountain, which is part of an anticline of Tsumeb Subgroup rocks, and substantial folding and thrusting can be observed within the gorge. Road D3704 to Opuwo turns to the N immediately after the gorge, and then transects dolomite of the Tsumeb Subgroup in a spectacular drive through small valleys and a steep pass. The road then crosses a peneplain underlain by limestones and dolomites of the Abendab Subgroup, and the red-brown, iron-rich Chuos Formation, base of the Tsumeb Subgroup, can be seen in places to the W of the road.

The Chuos Formation was extensively explored for iron in the 1950s and 1960s and a large low-grade resource of approximately 130 million tons of ore has been identified W of Kaoko Otavi. The iron ore body is a hematite-magnetite iron formation grading 36,8 % iron and 5,35 % manganese. To date, mining has not taken place due to the remoteness of the area and the grade of the ore, although it is interesting to note that the local Himba women use the powdered hematite to colour their bodies in bright red.

As the road continues, the higher mountains in the distance both to the E and the W are composed of dolomite and limestone of the Tsumeb Subgroup with Chuos Formation at the base. In the middle distance, in the E, the Kranzkuppen, Fahle Kuppen and Schwarzkuppen are composed of dolomite of the Abendab Subgroup, while quartzite of the Nosib Formation underlies the valley. Approximateluy 82 km from Sesfontein, the road parallels a major dry river bed, flanked by a number of large Baobab trees. The Baobab trees are rather unusual in this southern part of the Kaokoveld and are indicative of the shallow ground water available in the area.

Road D3705 is reached after about 100 km, and the route turns E towards Opuwo, and flat lying sediments of the Dwyka Formation, Karoo Sequence, and calcrete form the range to the SE. In the NW, the mountain range is formed by Abenab Subgroup dolomite. The road now follows a valley underlain by sediments of the Nosib Group, and some 15 km from the junction a river has carved small, but impressive erosional features out of the Cenozoic valley fill. Some of these resemble those found in the Monument Valley in Arizona, USA. Opuwo, reached 132 km after Sesfontein, is underlain by dolomites of the Abenab Subgroup.

Following road D 3700 to the N, more rocks of the Abenab Subgroup are exposed. Interlayered arkoses and dolomites form ridges, while softer siltstones are covered by calcrete and alluvials. About 10 km outside Opuwo, the road makes a marked turn to the W and then follows the valley of one of the tributaries of the Hoarusib River. This valley is underlain by tillite of the Dwyka Formation, and was formed by the Dwyka glaciation some 300 Ma ago. The Steilrandberge seen in the distance to the W are composed of Tsumeb Subgroup rocks.

Just S of Ohandungu, a thick package of dolomite breccia is exposed, and the contact between the Chuos Formation diamictite and the overlying carbonate rocks of the Tsumeb Subgroup is revealed in a traverse from the village of Ohandungu to the top of the Onjaraka Mountain SW of the road. Multiple cycles of carbonate debris flows are developed, and along the next ridge to the N, the contact between arkoses of the Nosib Group and the carbonate rocks of the overlying Abenab Subgroup are exposed.

Arkose, sandstone and siltstone of the Nosib Group form outcrops for the next 20 km, and are rather monotonous, except for of hematite pseudomorphs after pyrite in some arkoses. The road then passes a major fault, albeit with no expression in the landscape, and gneisses and amphibolites of the Epupa Complex are now exposed. The Epupa Complex is a succession of metamorphosed volcanic and sedimentary rocks, it is widely migmatised and altered to form porphyroblastic gneiss, feldspathic quartzite and biotite schist. It underlies large parts of northwestern Namibia, and, with an age of more than 2000 Ma, represents the oldest known rocks in the country.

The Epupa Complex is intruded by the slightly younger Kunene Complex, where anorthosite is the main rock type. Ultramafic to mafic, syenitic and carbonatitic intrusions of the Kunene Complex can be seen along road 3702, which turns off to the E at Epembe and leads to the sodalite mine at Swartbooisdrif. The rare, dark blue sodalite has been quarried for ornamental stone since 1964, and at present, the Swartbooisdrif mine is one of only two mines in the World producing sodalite. High-quality material is sold as semi-precious stone. The sodalite dykes are associated with and hosted by syenite, gabbro and anorthosite of the Kunene Complex.

Sporadic outcrops of gneisses of the Epupa Complex continue to Okangwati (Otjijanjasemo), until the Omuhonga Mountain N of Okangwati appears, formed by conglomerate of the Nosib Group, overlain by limestones of the Tsumeb Subgroup. Further to the E, the Zebra Mountains are an impressive range of anorthosite and troctolite, which are part of the Kunene Complex. The conspicuous stripes gave the range its name, and are not caused by igneous layering, but by alternating scree and more vegetated slopes.

Passing Omuhonga Mountain, the anorthosite and troctolite give way to the so-called "white anorthosite". The "white anorthosite" probably developed through potassic metasomatism and forms the bulk of the complex to the N, just S of Epupa. Within the "white anorthosite", massive magnetite-ilmenite plugs can be found in the complex, one of which occurs about 1 km S of the village of Ovizoreya. The road crosses this plug and shiny black magnetite-ilmenite rubble is abundant. The composition of the magnetite-ilmenite rock is variable, but on average has an iron content of about 55 % with 23 % titanium oxide and 0,5 % vanadium oxide.

Interestingly, a structural lineament, the NW-SE-trending Serpa Pinto Line, separates the dark grey to black anorthosites and troctolites from the bulk of the white anorthosite to the N, and may have controlled the metasomatism. Granites have also intruded the Kunene Complex and the gneisses of the Epupa Complex along the Serpa Pinto Line.

Some 4 km SW of Epupa Falls, the road makes a sharp turn to the E, and after another 2 km, a track leads towards a hill, which provides a magnificent bird's eye view of the falls. At the falls, the Kunene River has carved a narrow gorge into banded gneisses of the Epupa Complex along the contact with the harder, white anorthosite of the Kunene Complex. The falls, which mark the border between Angola and Namibia, drop approximately 30 m over gneiss of the Epupa Complex. Epupa is the Herero word for "falling waters", but before the falls, the river fans out to a width of 500 m and accomodates several islands. It then splits into several channels cutting through the gneiss over a width of about 460 m across the river. The falls themselves are a 1,5 km long stretch of tumbling cascades, and the main fall, the only one with a direct drop, is close to the southern bank and carries about one third of the water. The Kunene River valley is flanked by palm trees, while along the waterfalls Baobab trees also grow.

To proceed towards Marienfluss, one has to return to Okangwati (Otjijanjasemo), where road D3703 turns off to the W, and just W of the village, hot springs provide a pleasant surprise. These springs occur in gneisses of the Epupa Complex, which are cut in several places by basaltic dykes. From here the road continues westwards in a peneplain underlain by gneiss and granodiorite of the Epupa Complex, dissected by the occasional dry river bed. Towards the NW, the Otjihipa Mountains are prominent. They are composed of schist and quartzite of the Nosib Group overlain by limestone and dolomite of the Tsumeb Subgroup, Damara Sequence.

As Van Zyl's Pass is approached, a sudden change in the landscape marks the transition from the Epupa Complex to the rocks of the Okapuka Formation. The Okapuka Formation consists of a basal conglomerate, overlain by massive quartzite, feldspathic quartzite and phyllite, with towards the top amphibole schist and interbedded acid lava. This formation is interpreted to be an equivalent of the Khoabendus Group, and forms the eastern flanks of the Marienfluss valley. The flanks of this valley are extremely steep, and as the Van Zyl's Pass negotiates an altitude difference of more than 600 m in less than 8 km, the rugged topography caused by the massive quartzites and the intercalated acid volcanics make the road very difficult and advanced 4x4 driving skills are required to travel this pass.

Van Zyl's Pass leads directly into the wide valley of the Marienfluss (Otjihiipa) and in this valley, a wide, sandy plane with no outcrops extends for some 60 km. It ends at Otjinhungwa on the Kunene River, at the northern border of Namibia with Angola, and across the river, the Serra Cafema Mountains represent the northern extension of the rocks of the Okapuka Formation. The shallow river looks inviting, but crocodiles are a constant danger.

Travelling S again, the Hartmannberge with white marbles and schists of the Karibib and Kuiseb Formations of the Western Kaoko Zone cut by pegmatitic veins form the western fringe of the Marienfluss valley. Within the schist, a single layer with the rare orange mandarin garnet occurs, and has been mined intermittendly in the northern Marienfluss and in the Hartmannberge. Proceeding S, after approximately 35 km, the intersection between the tracks to Van Zyl's Pass and the Rooidrom (Red Drum) is reached and the route continues S and then turns to the W at the southern margin of the Marienfluss valley. The Rooidrom is some 71 km S of Otjinhungwa, and from here the road turns S to Orupembe. Marbles, quartzites and schists form the Otjihiipa Mountains to the E, and in the W, basalts of the Etendeka Plateau (see 8.6) occur. About 20 km S of the Rooidrom, a low mountain pass exposes marble, arkose, siltstone and sandstone of the Karibib Formation of the Damara Sequence.

Orupembe is reached 50 km S of the Rooidrom after crossing an area underlain by tillite of the Dwyka Formation, Karoo Sequence. The mountains to the N of Orupembe consist of lavas of the Etendeka Group, while the rugged mountains towards the E are once again composed of rocks of the Okapuka Formation. A waterhole drilled at Orupembe supplies the local villagers with water, and also springbok, oryx, kudu and giraffe.

Leaving Orupembe on road D 3707 in a southwesterly direction over flat gravel plains, the rugged mountains in the E are soon replaced by the typical table mountains formed by flat-lying flood basalts of the Etendeka Group with Karoo Sequence sediments at their base. Some 39 km S of Orupembe the road follows for a short while the bed of the Sechomib River, and thereafter it continues along the eastern fringes of a range of hills composed of schist of the Karibib Formation, Damara Sequence. At the northern end of this range, a spring sustains lush vegetation, and abundant artefacts and stone circles bear witness of early settlements. The Khumib River is crossed 26 km from here, with the series of Etendeka table montains still dominant in the E.

For the remaining 38 km to Purros, the road continues around the western and southern fringe of the Kabere, which, at 1105 m above sea level, is one of the highest Etendeka mountains in the area. At Purros, the Hoarusib River is crossed, and this river flows generally throughout the year on or near the surface. Spectacular cliffs of granitic gneisses of the Epupa Complex can be seen approximately 1 km downstream of the river crossing.

S of Purros, the road follows the Gomatum River towards Tomakas, and along the road steep cliffs expose marble and mica schist of the Kuiseb Formation, Damara Sequence, and gneiss of the Epupa Complex in a series of isoclinal to overturned synclines and anticlines. The high metamorphic grade is indicated by abundant migmatites, granitoids and pegmatites. The valley itself, however, is underlain by glacial sediments of the Dwyka Formation, Karoo Sequence, which indicates that this landform formed during the Dwyka glaciation.

The Purros Lineament, which separates the Western Kaoko Zone from the Central Kaoko Zone of the Damara Orogen, is crossed some 10 km from Purros. The road then leads into the Giribes Vlakte from Tomakas onwards, also underlain by the Dwyka Formation. E of Giribes Vlakte, some 40 km W of Sesfontein, the Epupa Complex is once again exposed. The rocks are greyish to brownish para-gneisses with sporadic granitic remobilisates in veins and stringers, and conspicuous augen-gneiss and migmatite become more common towards the W. Minor occurrences of amphibolite are also preserved. The Ganamub River, a major tributary to the Hoanib River, is crossed at the southeastern margin of Giribes Vlakte.

After crossing the river, Damaran rocks are once again exposed, and for a short while, the area is underlain by conglomerates of the Nosib Group. To the immediate W of Sesfontein, diamictites and ironstones of the Chuos Formation are traversed, before continuing in coarse- to medium-grained meta-arkose and quartzite of the Ugab Subgroup. This area is near the Sesfontein Thrust, a major tectonic lineament, and the rugged topography reflects the intense deformation. The road links up with a tributary of the Hoanib River after a steep descent, which it follows for 9 km, before continuing in the

Hoanib River valley. The mountains seen to the N and S are composed of schist of the Ugab Subgroup.

Some 2,5 km W of Sesfontein, a decrease in the metamorphic grade represents the change to a different tectono-metamorphic terrane. The boundary between the two terranes is represented by the Sesfontein Thrust constituting a series of W-dipping low-angle thrust planes. To the W, the meta-arkoses, quartzites, diamictites and ironstones represent a slope facies and have a higher metamorphic grade. To the E, the rocks of the Otavi Group comprise a thick package of white, pale, calcareous or dolomitic micrite, overlain by thinly bedded, grey, cherty dolomitic micrite with interlayered stromatolitic dolomites and dolomitic sedimentary breccias. These sediments are interpreted to represent a platform facies of the margin of the Congo craton.

The Sesfontein Thrust is actually crossed about 4 km W of Sesfontein. In the river bed it is concealed by the Cenozoic sediments of the Hoanib River, but it can be traced in the rocks on the northern bank, where the contact between the meta-arkose and quartzite of the Ugab Subgroup and the dolomite of the upper Tsumeb Subgroup is separated by a mylonitic zone. Along strike of the Sesfontein Thrust, a number of springs can be found, giving the name to Sesfontein, meaning “six fountains”. Sesfontein is reached 208 km S of Orupembe.

# 10. References

Anhaeusser, C.R. & Button, A. (1974): A review of southern African stratiform ore deposits – their position in time and space. – Inf. Circ. Econ. Geol. Res. Unit **85**, Univ. Witwatersrand, Johannesburg.

Bigella, J.J. (1970): Continental drift and palaeocurrent analysis (A comparison between Africa and South America). – Proc. Pap. IUGS 2nd Gondwana Symp., South Africa, pp. 73–97.

Buch, M.W. (1996): Geochrono-Geomorphostratigraphie der Etoscha Region, Nord-Namibia. – Die Erde **127**: 1–22.

Buch, M.W. & Rose, D. (1996): Mineralogy and geochemistry of the sediments of the Etosha Pan Region in northern Namibia: A reconstruction of the depositional environment. – J. Afr. Earth Sci. **22** (3): 355–378.

Bürg, G. (1942): Die nutzbaren Minerallagerstätten von Deutsch-Südwestafrika. – Mitt. Forschungsst. Kolonial. Bergbau Bergakad. Freiberg **2**, 305 pp.

Boise, C.W. (1915): Diamond Fields of “German” South West Africa I. – SA Mining J. **1242**: 468.

Cloos, H. & Chudoba, K. (1931): Der Brandberg. Bau, Bildung und Gestalt der Jungen Plutone in Südwestafrika. – N. Jahrb. Miner., Geol., Paläont., Beih. **66B**: 1–130.

Cloos, H. (1951): Gespräch mit der Erde – geologische Welt- und Lebensfahrt. – 389 pp., R. Piper & Co., München.

Corbett, I.B. (1989): The sedimentology of diamondiferous deflation deposits within the Sperrgebiet, Namibia. – 430 pp., unpubl. Ph.D. thesis, Univ. Cape Town.

Daltry, V.D.C. (1992): Type Mineralogy of Namibia. – Bull. Geol. Surv. Namibia **1**, 142 pp.

Deane, J. (1993): The controls on “contact-type” Cu-Pb(Ag) mineralisation within the Tsumeb Subgroup of the Otavi Valley syncline, northern Namibia. – 240 pp., M.Sc. Thesis (unpubl.), Univ. Cape Town.

Department of Water Affairs (1978): Grundwasser in Südwestafrika. – 39 pp., Windhoek.

de Waal, S. (1966): The Alberta Complex, a metamorphosed layered intrusion north of Nauchas, South West Africa, the surrounding granites and repeated folding in the younger Damara System. – D.Sc. thesis, Univ. Pretoria, 203 pp. (unpubl.).

de Wit, M.C.J. (1999) Post-Gondwana drainage and the development of diamond Placers in western South Africa. – Econ. Geol. **95**: 635–648.

Diehl, M. (1990): Geology, mineralogy, geochemistry and hydrothermal alteration of the Brandberg alkaline complex, Namibia. – Geol. Surv. Namibia Mem. **10**, 55 pp.

Diehl, B.J.M. & Schneider, G.I.C. (1990): Geology and mineralisation of the Rubicon Pegmatite, Namibia. – Open File Rep. **EG084**, 20 pp., Geol. Surv. Namibia.

du Toit, A.L. (1921): The Carboniferous glaciation of Southern Africa. – Trans. Geol. Soc. S. Afr. **24**: 188–227.

du Toit, A.L: (1927): A geological comparison of South America with South Africa. Carnegie Inst. Publ. **381**: 1–158, Washington D.C.

El Goresy, A. (1976): Opaque Oxide Minerals in Meteorites. – In: Rumble, D. (Ed.): Oxide Minerals. – Rev. Miner. **3**: EG47–71, Mineralogical Society of America, Washington.

Emmermann, R. (1979): Aufbau und Entstehung des Erongo Complexes. – In: Blümel, W.D., Emmermann, R. & Hüser, K. (Eds.): Der Erongo. Geowissenschaftliche Beschreibung und Deutung eines südwestafrikanischen Vulkankomplexes. – Sci. Res. SWA Ser. **16**: 16–53, Namibia Scientific Society, Windhoek.

Ewart, A., Milner, S.C., Armstrong, R.A. & Duncan, A.R. (1998): Etendeka Volcanism of the Goboboseb Mountains and Messum Igneous Complex, Namibia. Part I: Geochemical Evidence of Early Cretaceous Tristan Plume Melts and the Role of Crustal Contamination in the Parana-Etendeka Continental Flood Basalts. – J. Petrol. **39** (2): 191–226, Oxford University Press.

Galton, F. (1889): Narrative of an explorer in tropical South Africa. – Ward, Lock & Co., London.

Garvie, L.A.J., Devouard, B., Groy, T.L., Cemara, F. & Buseck, P.R. (1999): Crystal structure of kamenite, $NaHSi_2O_5$ x $3H_2O$, from Aris Phonolite, Namibia. – Amer. Mineral. **84**: 1170–1175.

Gebhard, G. (1991): Tsumeb – eine afrikanische Geschichte. – 239 pp., Verlag C. Gebhard, Giessen.

Genis, G. & Schalk, K.E.L. (1984): The geology of Area 2618: Keetmanshoop. Explanation of Sheet 2618, 1: 250 000. – Geol. Surv. SW Afr./Namibia, 12 pp., Windhoek.

Germs, G.J.B. (1972): New shelly fossils from the Nama Group, South West Africa. – Amer. J. Sci. **272**: 752–761.

Gerschütz, S. (1996): Geology, volcanology and petrogenesis of the Kalkrand Basalt Formation and the Keetmanshoop Dolerite Complex, southern Namibia. – 186 pp., Ph.D. thesis, Univ. Würzburg.

Gevers, T.W., Hart, O. & Martin, H. (1963): Thermal Waters Along the Swakop River, South West Africa. Trans. Geol. Soc. S. Afr. **66**: 157–189, Johannesburg.

Gührich, G. (1926): Über Saurier-Fährten aus dem Etjo-Sandstein von Südwestafrika. – Paläont. Z. **8** (1): 112–120.

Gührich, G. (1933): Die Kuibis-Fossilien der Nama-Formation von Südwestafrika. – Paläont. Z. **15**: 137–154.

Harris, C., Marsh, J.S. & Milner, S.C. (1999): Petrology of the Alkaline Core of the Messum Igneous Complex, Namibia: Evidence for the Progressively Decreasing Effect of Crustal Contamination. – J. Petrol. **40** (9): 1377–1398, Oxford University Press.

Hawthorne, J.B. (1975): Model of a kimberlite pipe. – Phys. Chem. Earth **9**: 1–15.

Hedberg, R.M. (1979): Stratigraphy of the Owamboland basin, South West Africa. – Bull. Precambrian Res. Unit 24, 325 pp., University Cape Town, Cape Town. [Ph.D. thesis].

Hegenberger, W. (1987): Stand der geologischen Kenntnisse über das Kavangogebiet. – J. SWA Sci. Soc. **XLI**: 97–113, Windhoek.

Heine, K. (1992): On the ages of humid Late Quaternary phases in southern African arid areas (Namibia, Botswana). – Palaeoecol. Afr. Surround. Islands **23**: 149–164.

Hellwig, D.H.R. (1978): Evaporation of water from sand 6: The influence of the depth of the water-table on diurnal variations. – J. Hydrol. **39**: 129–138.

Hoad, N. (1992): Resource assessment and development potential of contrasting aquifers in Central Northern Namibia. – Unpubl. M.Sc. thesis, 67 pp., Univ. College London.

Hoad, N. (1993): Resource assessment and development potential of contrasting aquifers in central northern Namibia. – Ph.D. thesis, 67 pp., Univ. College London.

Hoad, N. (1993): An overview of groundwater investigations in the Tsumeb and Oshivelo areas – Unpubl. rep., 64 pp., file no.: 12/4/2/18, Department of Water Affairs, Windhoek.

Hoffman, P.F., Kaufmann, A.J. & Halverson, G.P. (1998a): Comings and goings of global glaciations on a Neoproterozoic tropical platform in Namibia. – GSA Today **8** (5): 1–9.

Hoffman, P.F., Kaufmann, A.J., Halverson, P.G. & Schrag, D.P. (1998b): A Neoproterozoic Snowball Earth. – Science **281**: 1342–1346.

Hunter, D.R. & Pretorius, D.A. (1981): Structural Framework. – In: Hunter, D.R. (Ed.): Precambrian of the Southern Hemisphere. – pp. 397–419, Elsevier, Amsterdam.

Ingwersen, G. (1990): Die sekundären Mineralbildungen der Pb-Zn-Cu Lagerstätte Tsumeb, Namibia. – Ph.D. thesis, 233 pp., Univ. Stuttgart.

Innes, J. & Chaplin, R.C. (1986): Ore bodies of the Kombat Mine, South West Africa/Namibia. – In: Annhaeusser, C.R. & Maske, S. (Eds.): Mineral Deposits of Southern Africa. – Spec. Publ. Geol. Soc. S. Afr. **2**. 1789–1805.

Jacob, G. (1987): Reise zur Dioptas Mine Omaue in Kaokoveld. – Lapis **12** (4): 11–14.

Jacob, R.J., Bluck, B.J. & Ward, J.D. (1999): Tertiary-age diamondifeous fluvial deposits of the lower Orange River valley, southwestern Africa. – Econ. Geol. **95**: 749–758.

Jaeger, F. & Waible, F. (1920): Beiträge zur Landeskunde von Südwestafrika. – Mitt. deutsch. Schutzgeb., Erg.-H. **14**, 80 pp., Berlin.

Kaiser, E. (1912): Ein neues Beryll(Aquamarin)-Vorkommen in Deutsch Südwestafrika. – Zentralbl. Mineral. Geol. Paläont. **13**: 385–390.

Kelber, K.-P., Franz, L., Stachel, T., Lorenz, V. & Okrusch, M. (1993): Plant fossils from Gross Brukkaros (Namibia) and their biostratigraphical significance. – Comm. Geol. Surv. Namibia **8**: 57–66.

Keller, P. (1977): Paragenesis. – Mineral. Rec. **8** (3): 38 – 47.

Keller, P. (1984): Tsumeb/Namibia – eine der spektakulärsten Mineralfundstätten der Erde. – Lapis **9** (7&8): 13 – 63.

Keller, P. (1991): The occurrence of Li-Mn-Fe phosphate minerals in granitic pegmatites of Namibia. – Comm. Geol. Surv. Namibia **7**: 21 – 34.

Keller, P. & von Knorring, O. (1989): Pegmatites at the Okatjimukuju farm, Karibib, Namibia. Part I: Phosphate mineral associations of the Clementine II pegmatite. – Eur. J. Mineral. **1**: 567 – 593.

Keller, P., Fontan, F. & Fransolet, A.-M. (1994): Intercrystalline cation partitioning between minerals of the triplite-zwieselite-magniotriplite and triphyllite-lithiophilite series in granitic pegmatites. Contr. Mineral. Petrol. **118**: 239 – 248.

Killick, A.M. (1986): A review of the economic geology of northern South West Africa/Namibia. – In: Annhaeusser, C.R. & Maske, S. (Eds.): Mineral Deposits of Southern Africa. – Spec. Publ. Geol. Soc. S. Afr. **2**: 1709 – 1717.

Klinger, E. (1977): Cretaceous deposits near Bogenfels, South West Africa. – Ann. S. Afr. Mus. **7** (3): 81 – 92.

Kok, T.S. (1964): The occurrence and location of groundwater in South West Africa. – Ph.D. thesis, 265 pp., Univ. Cape Town.

Korn, H. & Martin, H. (1954): The Messum Igneous Complex in South-West Africa. – Trans. Geol. Soc. S. Afr. **57**: 83 – 124.

Korn, H. & Martin, H. (1959): Gravity Tectonics in the Naukluft Mountains of South West Africa. – Bull. Geol. Soc. Amer. **70** (8): 1047 – 1078.

Kräusel, R. (1928): Fossile Pflanzenreste aus der Karruformation Deutsch Südwestafrikas. – Beitr. geol. Erforsch. deutsch. Schutzgeb. **20**: 17 – 54, Berlin.

Kräusel, R. (1956): Der "Versteinerte Wald" im Kaokoveld, Südwest-Afrika. Senckenberg. Lethaia **37** (5/6): 411 – 437, Frankfurt/M.

Kurzlaukis, S. (1994): Geology and geochemistry of the carbonatitic Gross Brukkaros Volcanic Field and the ultrabasic Blue Hills Intrusive Complex, southern Namibia. – Ph.D. thesis, 290 pp., Univ. Würzburg.

Kurzlaukis, S. & Lorenz, V. (1997): Volcanology of a low viscosity melt: the Carbonatitic Gross Brukkaros Volcanic Field, Namibia. – Bull. Volcanology **58**: 421 – 431.

Levinson, O. (1983): Diamonds in the Desert. – 172 pp., Tafelberg Publishers, Cape Town.

Lockley, M. (1991): Tracking Dinosaurs. – 238 pp, Cambridge University Press, Cambridge.

Limbaard, A.F., Günzel, A., Innes, J. & Krüger, T.L. (1986): The Tsumeb lead-copper-zinc-silver deposit in South West Africa/Namibia. – In: Annhaeusser, C.R. & Maske, S. (Eds.): Mineral Deposits of Southern Africa. – Spec. Publ. Geol. Soc. S. Afr. **2**: 1761 – 1787.

Mabbutt, J.A. (1950): The evolution of the middle Ugab valley, Damaraland, South West Africa. – Rep. Trans. Roy. Soc. S.Afr. **XXXIII**: 333 – 367.

Marais, J.C.E., Irish, J. & Martini, J.E.J. (1996): Cave investigations in Namibia V: 1993 Swakno results. – Bull. SA Spel. Ass. **36**: 58 – 78.

Marchant, J.W. (1980): Hydrogeochemical exploration at Tsumeb. – Unpubl. Ph.D. thesis, 222 pp., Cape Town University.

Marsh, A. & Seely, M. (1992): Oshanas. – 52 pp., DRFN, Windhoek.

Martin, H. (1963): A suggested theory for the origin and a brief description of some gypsum deposits of South West Africa. – Trans. Geol. Soc. S. Afr. **66**: 345 – 350.

Martin, H. (1970): The Sheltering Desert (German Original Title: Wenn es Krieg gibt, gehen wir in die Wüste). – 244 pp., Verlag der S.W.A. Wissenschaftlichen Gesellschaft, Windhoek.

Martin, H., Porada, H. & Wittig, R. (1983): The Root Zone of the Naukluft Nappe Complex: Geodynamic Implications. – In: Martin, H. & Eder, F.W. (Eds.): Intracontinental Folds Belts. – 679 – 698, Springer, Heidelberg.

Martini, J.E.J. (1992): Two new minerals originated from bat guano combustion in Arnhem Cave. – Bull. S.A. Spel. Ass. **33**.

Marvin, U.B. (1999): Historical Notes on three exceptional Meteorites of Southern Africa. – In: Proceedings of the Workshop on Extraterrestrial Materials from Hot and Cold Deserts. – pp. 48 – 52, Johannesburg.

Mc Corkell, R.H., Fireman, E.L. & d'Amico, J. (1968): Radioactive Isotopes in Hoba West and other Iron Meteorites. – Meteoritics **4** (2): 113 – 122.

Menge, G.F.W. (1986): Sodalite carbonatite deposits of Swartbooisdrif, SWA/Namibia. In: Annhaeusser, C.R. & Maske, S. (Eds.): Mineral Deposits of Southern Africa. – Spec. Publ. Geol. Soc. S. Afr. **2**: 2261 – 2268.

Miller, R. McG. (1979): The Okahandja lineament, a fundamental tectonic boundary in the Damara Orogen of South West Africa/Namibia. – Trans. Geol. Soc. S. Afr. **82**: 349 – 361.

Miller, R. McG. (1983): The Pan-African Damara Orogen of South West Africa/ Namibia. – In: McG. Miller, R. (Ed.): The Evolution of the Damara Orogen of South West Africa/Namibia. – Geol. Soc. S. Afr., Spec. Publ. **11**: 431 – 515.

Miller, R. McG. (1990): Damara Excursion. – Geocongress 1990, 116 pp, Geol. Soc. S. Afr., Johannesburg.

Miller, R. McG. (1992): Stratigraphy. – In: The Mineral Resources of Namibia. – 16 pp., Geological Survey of Namibia, Windhoek.

Miller, R. McG. (1997): The Owambo Basin of Northern Namibia. – In: Hsü, K.J. (Ed.): Sedimentary basins of the World: African Basins. – pp. 237 – 268, Elsevier, Amsterdam.

Miller, R.McG., Fernandes, L.M. & Hoffmann, K.H. (1990): The Story of Mukorob. – Journal **42**: 63 – 73, Namibia Wissenschaftliche Gesellschaft, Windhoek.

Milner, S.C. (1986): The Geological and Volcanological Features of the Quartz Latites of the Etendeka Formation. – Comm. Geol. Surv. SW Africa/Namibia **2**: 109 – 116.

Milner, S.C., Duncan, A.R., Marsh, J.S. & Miller, R. McG. (1988): Field Excursion Guide to the Etendeka Volcanics and Associated Intrusives, NW Namibia. – 51 pp, Windhoek.

Milner, S.C. & Ewart, A. (1989): The Geology of the Goboboseb Mountain Volcanics and their relationship to the Messum Complex, Namibia. – Comm. Geol. Surv. Namibia **5**: 31–40, Windhoek.

Milner, S. & Swart, R. (1994): Messum – Albin Excursion. – 16 pp. (unpubl.), Excursion Guide Geological Society of Namibia, 16, Windhoek.

Milner, S.C., Leroux, A.P. & O'Connor, J.MJ (1995): Age of Mesozoic Igneous Intrusions in Northwestern Namibia, and their Relationship to Continental Break-up. – J. Geol. Soc. London **151**: 97–104.

Ministry of Agriculture, Water and Rural Development (1993): Central Area Water Masterplan: Phase 1, 11. – 64 pp, Windhoek.

Münch, H.G. (1974): The Tectonics of the Northern Part of the Klein Karas and Groot Karas Mountains, South West Africa. – Ann. Geol. Surv. S. Afr. **9**: 107–109.

Netterberg, F. (1980): Geology of Southern African Calcretes: 1. Terminology, Description, Macrofeatures and Classification. – Trans. Geol. Soc. S. Afr. **83**: 255–283.

Norton, O.R. (1998): Rocks from Space. – 449 pp., Mountain Press Publishing Company, Missoula, Montana.

Penney, A.J., Maxwell, C.D. & Roux, D.E. (1988): Guinas Lake. – Bull. S. Afric. Spelaeol. Assoc. **29**: 6–9.

Penrith, M.J. (1978): Otjikoto Lake. – South West Africa Annual **1978**: 138–139.

Pickford, M.F.H. (1994): A new species of *Prohyrax* (Mammalia, Hyracoidea) from the middle Miocene of Arrisdrift, Namibia. – Comm. geol. Surv. Namibia **9**: 43–62.

Pickford, M.F.H. (1996): Fossil crocodiles *(Crocodylus lloydi)* from the lower and Middle Miocene of southern Africa. – Ann. Paléont. **82**: 235–250.

Pirajno, F. (1990): Geology, geochemistry and mineralisation of the Erongo Volcanic Complex, Namibia. – S. Afr. J. Geol. **93** (3): 485–504.

Pirajno, F. & Schlögl, H.U. (1987): The alteration-mineralisation of the Krantzberg tungsten deposit, South West Africa/Namibia. – S. Afr. J. Geol. **90**: 499–508.

Pirajno, F., Kinnaird, J.A., Fallick, A.E., Boyce, A.J. & Petzel, V.F.W. (1993): A preliminary regional sulphur isotope study of selected samples from mineralized deposits of the Damara Orogen, Namibia. Comm. – Geol. Surv. Namibia **8**: 81–97.

Ploethner, D., Schmidt, G., Kehrberg, S. & Geyh, M.A. (1997): German Namibian Groundwater Exploration Project: Hydrogeology of the Otavi Mountain Land and its Surroundings (Karst_01 and Karst_02). – Unpubl. report, Vol. **D-III**, 65 pp., Bundesanstalt für Geowissenschaften und Rohstoffe, Hannover.

Porada, H.R. (1974): The Khoabendus Formation in the area northwest of Kamanjab and in the southeastern Kaokoland, South West Africa. – Mem. Geol. Surv. S. Afr., S.W. Afr. Ser. **4**, 23pp.

Porada, H. & Wittig, R. (1976): Das Chausib-Turbiditbecken am Südrand des Damara-Orogens, Südwestafrika. – Geol. Rdsch. **65**: 1002–1019.

Reeves, C.V. (1972): Rifting in the Kalahari? – Nature **237** n, 5350, 95–96.

Reid, D.L. (1998): Karoo Dolerite in a Kokerboom Forest. – In: Reid, D.L. (Ed.): Field Excursion Guide. – IAVCEI, 6–15, Cape Town.

Reuning, E. & Martin, H. (1957): Die Prä-Karoo-Landschaft, die Karoo-Sedimente und Karoo-Eruptivgesteine des südlichen Kaokoveldes in Südwestafrika. – N. Jb. Miner. Geol. Paläont. **91**: 193–212.

Richards, T.E. (1986): Geological characteristics of rare metal bearing pegmatites of the Uis type in the Damara Orogen, South West Africa/Namibia. – In: Annhausser, C.R. & Maske, S. (Eds.): Mineral Deposits of Southern Africa. – Geol. Soc. S. Afr., Spec. Publ. **2**: 1845–1862.

Rickwood, P.C. (1981): The largest crystals. – Amer. Miner. **66**: 885–907.

Rose, D. (1981): Multi-step emplacement of a pegmatitic vein – Brabant pegmatite, Namibia. – N. Jb. Mineral. Mh. **1981**: 355–373.

Runnegar, B. & Fedonkin, M. (1992): Proterozoic Metazoan body fossils. – In: Schopf, J.W. & Klein, C. (Eds.): The Proterozoic Biosphere – A multidisciplinary study. – pp. 369–388, Cambridge University Press.

Schalk, K.E.L. (1982): Geologische Geschichte des Gamsberggebietes. – J. SWA Wiss. Ges. **XXXVIII**: 7–15, Windhoek.

Schalk, K.E.L. & Germs, G.J.B. (1980): The Geology of the Mariental Area. Explanation of Sheet 2416 Mariental. – 7 pp., Geol. Surv. Namibia, Windhoek.

Schmidt, G. & Plöthner, D. (1999): Grundwasserstudie zum Otavi Bergland/Namibia. – Z. Angew. Geol. **45**: 114–123.

Schmidt-Thome, P. (1981): Ist der Fischfluß-Canon in Südwestafrika (Namibia) durch eine Grabenstruktur vorgezeichnet? – Geol. Rdsch. **70** (2): 499–503.

Schmitt, A.K., Emmermann, R., Trumbull, R.B., Bühn, B. & Henjes Kunst, F. (in prep.): Petrogenesis and Ar/Ar Geochronology of Metaluminous and Peralkaline Granites from the Brandberg Complex, Namibia: Evidence for major mantle contribution.

Schneider, G.I.C. (1992): Manganese. – In: The Mineral Resources of Namibia. – Geol. Surv. Namibia, Spec. Publ., p 2.6-1–2.6-9.

Schneider, G.I.C. & Miller, R. McG. (1992): Diamonds. – In: The Mineral Resources of Namibia. – Geol. Surv. Namibia, Windhoek, Spec. Publ., p. 5.1-1–5.1-32.

Schneider, G.I.C. & Seeger, K.G. (1992): Semi-precious stones. – In: The Mineral Resources of Namibia. – Geol. Surv. Namibia, Spec. Publ., p. 5.2-1–5.2-16.

Scholz, C.H. (1976): Rifting in the Okavango Delta. – Natural History Magazine **85** (2): 38–42.

Scott, L., Cooremans, B., de Wet, J.S. & Vogel, J.C. (1991): Holocene environmental changes in Namibia inferred from pollen analysis of swamp and lake deposits. – The Holocene **1**: 8–13.

Seeger, K.G. (1990): An evaluation of the groundwater resources in the Grootfontein Karst area. – Unpubl. rep., 187 pp., Department of Water Affairs, Windhoek.

Seth, B. (1998): Archaean and Neoproterozoic magmatic events in the Kaokobelt of northwestern Namibia, and their geodynamic significance. – Precambr. Res. **92**, 341–363.

Shepard, C.U. (1853): Notice of Meteoric Iron near Lion River, Great Namaqualand, South Africa. – Amer. J. Sci., 2nd Ser., **XV**: 1–4.

Skelton, P.H. (1990): The Status of Fishes from Sinkholes and Caves in Namibia. – Bull. S Afric. Spelaeol. Assoc. **31**: 77 – 80.

Smith, A.M. & Mason, T.R. (1991): Pleistocene, multiple-growth, lacustrine oncoids from the Poacher's Pan Point Formation, Etosha Pan, northern Namibia. – Sedimentology **38**: 591 – 599.

Smith, R. & Swart, R. (2000): Field Guide to the rocks and fossils of Etjo Mountain. – 10 pp, Geological Society of Namibia, Windhoek.

Stachel, T., Brey, G. & Stanistreet, I.G. (1994): Gross Brukkaros (Namibia) – The unusual intra-caldera sediments and their magmatic components. – Comm. Geol. Surv. Namibia **9**: 23 – 42.

Stachel, T., Brey, G. & Lorenz, V. (1995): Carbonatite magmatism and fenitization of the epiclastic caldera-fill at Gross Brukkaros (Namibia). – Bull. Volcanol. **57**: 185 – 196.

Stengel, H.W. (1966): The rivers of the Namib and their discharge into the Atlantic. Part II: Omaruru and Ugab. – Sci. pap. Namib Desert Res. Stat. **30**, 35 pp.

Stengel, H.W. (1968): Wasserspeicherung in den Sanden eines Riviers. – Wiss. Forsch. S.W. Afrika **7**, 54 pp., S.W.Afrika Wiss. Ges., Windhoek.

Steven, N.M. (1993): A study of the epigenetic mineralisation in the Central Zone of the Damara Orogen, with special reference to gold, tungsten, tin and rare earth elements. – Mem. Geol. Surv. Namibia **16**, 166 pp.

Stewart, W.N. & Rothwell, G.W. (1983): Palaeobotany and the Evolution of Plants. – 521 pp, Cambridge University Press, New York.

Symons, G., Miller, R.McG., Hata, Y., Yamasaki, Y. & Noell, U. (2000): Geology and resistivity characteristics of the Stampriet Artesian Basin. – In: Sililo, O. et al. (eds.): Proceedings of the International Association of Hydrologists, XXX Congress on Groundwater. Cape Town.

Thirion, N.C. (1970): Outjo Townlands Grant M46/3/298. – Unpubl. Report, pp. 14, Tsumeb Corporation Limited.

Tordiffe, E.A.W. (1996): Enhanced groundwater recharge tests on the Omdel aquifer in Namibia. – Workshop on Groundwater-Surface water issues in arid and semi-arid areas, 17 pp., Warmbad.

Tredoux, G. & Kirchner, J. (1980): The evolution of the chemical composition of the artesian water in the Auob Sandstone. – Groundwater 1980, 8 pp, Pretoria.

Tredoux, G. & Kirchner, J. (1981): The evolution of the chemical composition of the artesian water in the Auob Sandstone. – Trans. Geol. Soc. S. Afr. **84**: 169 – 175.

Tredoux, G. & Kirchner, J. (1985): The occurrence of nitrate in groundwater in South West Africa/Namibia. – Conference Proceedings, 2, Nitrates dans les Eaux, Paris.

Trewavas, E. (1936): Dr. Karl Jordan's expedition to South-West Africa and Angola: The Freshwater Fishes. – Novitates Zoologicae **40**: 63 – 74.

Trompette, R. (1991): Geology of Western Gondwana. – 350 pp, Balkema, Rotterdam.

Trumbull, R.B., Emmermann, R., Bühn, B., Gerstenberger, H., Mingram, B., Schmitt, A. & Volker, F. (2001): Insights on the genesis of the Cretaceous Damaraland igneous complexes in Namibia: The isotopic perspective. – Comm. Geol. Surv. Namibia **12**: 313–324, Windhoek.

Truswell, J.F. (1977): The Geological Evolution of South Africa. – 218 pp, Purnell, Cape Town.

Uebel, P.-J. (1977): Internal structure of pegmatites, their origin and nomenclature. – N. Jahrb. Mineral., Abh. **131**: 83–113.

Vogel, J.C. & van Urk, H. (1975): Isotopic composition of groundwater in semi arid regions of southern Africa. – J. Hydrol. **25** (2): 23–36.

Wagner, P.A. (1914): The diamond fields of Southern Africa. – C Struik, Cape Town, reprint 1971, 355 pp.

Ward, J. & Jacob, J. (1998): The Orange River, provider of many facets. – EWI Namdeb 1 (1998): 8–11.

Ward, J.D., Seely, M.K. & Lancaster, N. (1983): On the antiquity of the Namib. – S. Afr. J. Sci. **79**: 175–183.

Whitten, D.G.A. & Brooks, J.R.V. (1981): The Penguin Dictionary of Geology. – 495 pp., Penguin Books, Harmondsworth.

Wilson, W.E. (1977): Tsumeb! The World´s greatest Mineral Locality. – Mineral. Rec. **8** (3), 128 pp.

Wittig, R. (1976): Die Gamsberg-Spalten – Zeugen Karoo-zeitlicher Erdbeben. – Geol. Rdsch. **65** (3): 1019–1034.

# 11. Index

€ 38,00